Factory-Constructed Housing Developments

Planning, Design, and Construction

Factory-Constructed Housing Developments

Planning, Design, and Construction

William F. Albern, P.E.

Edited by
M.D. Morris, P.E.

All photographs by William F. Albern, P.E.

CRC Press

Boca Raton   New York

Library of Congress Cataloging-in-Publication Data

Albern, William F.
    Factory-constructed housing developments : planning, design, and
construction / William F. Albern.
        p. cm.
    Includes index.
    ISBN 0-8493-7481-2 (alk. paper)
    1. Prefabricated houses. 2. Industrialized building. 3. Housing
development. I. Morris, M. D. (Morton Dan) II. Title.
TH4819.P7A56 1997
711'.58—dc21

96-29774
CIP

This book contains information obtained from authentic and highly regarded sources. Reprinted material is quoted with permission, and sources are indicated. A wide variety of references are listed. Reasonable efforts have been made to publish reliable data and information, but the author and the publisher cannot assume responsibility for the validity of all materials or for the consequences of their use.

Neither this book nor any part may be reproduced or transmitted in any form or by any means, electronic or mechanical, including photocopying, microfilming, and recording, or by any information storage or retrieval system, without prior permission in writing from the publisher.

The consent of CRC Press LLC does not extend to copying for general distribution, for promotion, for creating new works, or for resale. Specific permission must be obtained in writing from CRC Press LLC for such copying.

Direct all inquiries to CRC Press LLC, 2000 Corporate Blvd., N.W., Boca Raton, Florida 33431.

No claim to original U.S. Government works
International Standard Book Number 0-8493-7481-2
Library of Congress Card Number 96-29774
Printed in the United States of America 1 2 3 4 5 6 7 8 9 0
Printed on acid-free paper

Preface

This text does not include everything you need to know to design a manufactured home community or small subdivision. However, together with the cited references, you will find the task achievable. My primary intent has been to assemble or cite the extensive available literature, add to it the little details learned over the years, and make the information accessible to others who may be embarking on a career to assist owners with the creation of new developments. The appendices contain information that will be of interest to many. Municipal boards should find several items that can provide ideas for their operations. Manufactured home community operators will find other useful items. I hope they, and the designers of developments, will find much useful information.

Living is better in an attractive development.

William F. Albern, P.E.
Ithaca, New York

About the Author

William F. Albern, P.E., specializes in two distinctive areas of engineering design. His original training and specialty is in the area of heating, ventilation, air conditioning, plumbing, and process piping. Due to a work environment which presented a unique opportunity to acquire experience, he also provides engineering services in the area of factory-constructed housing and subdivision developments. In his professional career he has been associated with major engineering manufacturers and consulting firms, including a long association with Energy Management for Facilities Engineering at Cornell University.

Mr. Albern is a graduate of Clarkson University, Potsdam, NY, where he received his Bachelors of Mechanical Engineering in 1951. His technical society affiliations have included the American Society of Heating, Refrigerating and Air-Conditioning Engineers (ASHRAE), the American Society of Plumbing Engineers (ASPE), the Construction Specifications Institute, and the National Society of Professional Engineers.

He was the first President of the Upstate New York Chapter of ASPE in 1972 and also the first President of the Twin Tiers Chapter of ASHRAE in 1977. Subsequently, he served ASPE as the Secretary of the Society for six years and served ASHRAE as a Regional Vice Chairman, Regional Chairman, and member of the Society's Board of Directors. He has been elected a Fellow of ASHRAE.

Mr. Albern was the Engineer of the Year in 1986 for the Broome Chapter, New York State Society of Professional Engineers, and is listed in *Who's Who in Technology, 6th Edition,* 1989.

Acknowledgments

The first thank you must go to Dan Morris, the editor of this epistle. Without Dan's perseverance it would never have come about. He insisted the subject was large enough, and that an audience existed.

To Jim Ray, Sr., a client since 1975, a friend since shortly thereafter, and the contributor of the first chapter, I say thank you for all you have done for me for the past 20 years. It has been a pleasure working with you.

Jim "Chip" Ray, Jr., A.C.M., thank you for the enlightenment on the Accredited Community Manager program for manufactured home communities.

Brayton Foster, a consulting geologist and the other contributor, has been an acquaintance for only a few years. My understanding of groundwater resources, and wells in particular, has increased immeasurably as a result.

Craig White, A.C.M., and his High Meadow, factory built home, realty subdivision in Florence, CO, added a dimension to the book.

Harold Payton, who appears in Figure 6.1, took extra time on a project to allow me to photograph well drilling equipment.

Michael Hovanec and his wife, Anna, and Paul Jacobs were most helpful with some of the contacts with the manufacturers of factory-constructed housing.

I thank Alan Froehle, General Manager, Crest Homes, for his generosity helping me photograph the factory housing being constructed in his plant.

And, of course, my wife, Joan. How many times did she sit patiently (maybe it was impatiently) in the car while I wandered through a subdivision or manufactured home community? And for the innumerable visits to factory-constructed housing and the muddy sites. Thanks, my love, neither one of us ever thought we would produce a book.

Table of Contents

Economics

1

JIM RAY, SR.

General

The economics of building and/or operating a manufactured home community will vary considerably from one region to another. Before getting deeply involved in the purchase or construction of a manufactured home community, you should hire an accounting firm experienced in the operation of a manufactured housing community who is capable of not only setting up a proper set of books, but who can do economic projections of the return on your investment. Many projects have been financial disasters because this preliminary economic work was not done.

The structure of ownership of the operation needs to be merged with your overall financial plan. Since the value of a large manufactured home community will be significant, it is essential that you choose a business structure that fits your situation. A corporation, a family partnership, a limited liability company, and sole proprietorship are only a few of the alternatives you can choose.

In 1958, at age 19, Jim Ray sold his first mobile home, a 35' × 10' Nashua, for about $3,000. With available spaces for new homes scarce, existing park owners were demanding under the table cash payments to permit a new home to enter their parks. Sensing the opportunity, Ray launched a search for a site for his own manufactured home community. Using a new, unique quadrangle design instead of the traditional "soldier in line" approach, his first park opened in 1960. The organization has now grown to four parks and 465 home sites.

Construction costs vary not only with the labor and material market and availability, but with the climate. All proposed new construction should enlist the help of the most experienced contractors and engineers to determine the most precise construction cost for those specific areas.

Fill rates will vary depending on the local economic conditions. No project should be started unless the market conditions warrant additional demand that is not being filled. Such market studies can be somewhat complex, and if you are not able to supervise and participate in the study yourself, or with a trusted professional familiar with that particular market, you could find yourself in financial trouble.

Filling a manufactured home community can be done in several ways. One is to let the existing dealers sell into your community. If they have an interest in another community, you can be sure they will direct their sales there. You could get a dealer to set up models in your community and even plan a sales center as part of your operation. This could be operated by another dealer or you could go into the sales business yourself. If you are able to do the selling yourself, you will find your fill rate will be faster and you can get back some of the investment in the project from the profit from the home sale.

Build according to the concepts suggested in this text.

Once you have begun operation of the community, it is absolutely essential to your financial health to keep accurate records of every aspect of the community's operation. Spreadsheets may serve to keep you abreast of key information and provide it to you for your review with little on-going time commitment. You can develop spreadsheets that will furnish the information you want in a way that is meaningful to you.

Some parts of the country assess the homes and communities on their real market value while others may use income valuation methods. There are wide differences as to the percentage of income that goes to real estate taxes. You need to familiarize yourself with the assessment procedures used in your area to determine the value to be placed on your community, since real estate taxes can be as high as 40% of your gross revenue in some areas.

I believe there are still opportunities for growth in this industry, so long as you choose an area where the market has not matured, and has enough demand for you to build and fill your project in a reasonable time.

One source of information about the "real world" of operating a manufactured home community is the state trade association for the industry. Most community owners like to talk about their operations and time spent at association meetings can be worthwhile. The associations usually have much valuable information to help both new and experienced owners remain aware of the issues that affect them, from zoning to state and local legislative initiatives by the association, and most important, by tenant and consumer organizations that tend to promote legislation that goes far beyond what is needed to solve problems. Whenever such legislation is passed, property owner rights are further eroded.

You will require administrative forms and spreadsheets to keep pertinent information available in a format that is meaningful to you. There is no set standard for each situation that occurs in the jurisdictions where manufactured communities are. It is the owners' responsibility to inform themselves of those areas of concern, and to take the administrative action necessary to develop a process that collects facts and puts them in an understandable form. Your accounting firm can be of great assistance with the preparation of information; just be sure you can understand it in time to take any required actions.

Be positive, you'll succeed.

Factory-Constructed Housing

2

Factory-Constructed Housing

Factory-constructed housing accounts for about 25% of single family homes constructed annually in the U.S. as stated in the U.S. Department of Commerce, C25 Construction Reports. The Department of Housing and Urban Development (HUD) sets the standards for the construction of manufactured housing.

The American Planning Association has stated:

> Since 1976, The U.S. Department of Housing and Urban Development has regulated certain types of factory-built housing under the Manufactured Home Construction and Safety Standards (hereinafter HUD Code). At that time, these housing units were called "mobile homes". In 1980, this designation was officially changed to "manufactured homes". This change was made, for the most part, in recognition of the more durable and less mobile nature of these modern factory-built units. Today's manufactured homes are still built on a wheeled chassis (they are the only factory-built units constructed in this manner, as required by the HUD Code), but, once they are sited, they are rarely moved, and the hitch, axles, and wheels are removed.*

So, today, the Federal government recognizes a home constructed in a factory with a wheeled chassis as a manufactured home. Some state governments are also moving in that direction. And, the state manufactured housing associations are using that language. To those entities, a mobile home is one that was constructed prior to the enactment of the HUD Code in 1976. Admittedly, the general public has yet to accept that terminology.

---

* Welford Sanders, Regulating Manufactured Housing, Planning Advisory Service Report No. 398, Chicago: American Planning Association, December 1986.

What is a Factory-Constructed Home?

What is a factory-constructed home? What is a manufactured home? What is a modular home? What is a mobile home? Modular homes and manufactured homes are factory-constructed homes. They are factory-built on some semblance of an assembly line. They are designed for year-around occupancy and contain all piping and wiring for connection to all necessary utilities at the site. As stated above, the terms have been changing over the years. To clarify those terms, at least as used within this text, I offer the following.

Modular homes are supported by a perimeter foundation that may form a crawl space or may be a full basement, or they may be supported by a concrete slab. A modular home may be of any configuration, and may be more than one story high. They are transported on trailers as would other types of freight. The trailer may be a unit specifically developed by the home manufacturer to transport its products. Modular homes are normally placed on land owned by the homeowner. A homeowner would probably have great difficulty securing a mortgage for a modular home placed on rented land.

Generally, modular homes must meet the requirements of the building construction code of the local community. This significantly raises the cost of modular housing. Each home needs to be custom constructed to meet the requirements of the jurisdiction where it will be placed. Manufacturers of modular housing cannot economically construct a home acceptable in all jurisdictions. This contrasts with the manufactured home industry constructing homes in accordance with the HUD Code that effectively is a uniform construction code acceptable throughout the country. An architect, frequently part of the construction of a modular home, is seldom, if ever, part of the siting of a manufactured home.

To provide more affordable housing for its citizens, some of the progressive states have developed a statewide modular housing code. These codes set the standard requirements for all jurisdictions in the state. The concept is similar to the HUD Code except it applies to modular housing on a state level. It prevents the arbitrary interpretation of individual county codes and thereby reduces the overall cost of housing in the state. I understand several of our western states have banded together and have a single multi-state modular code, another opportunity to control housing costs. Modular and manufactured homes can be constructed in a factory in a one- to two-week time frame from start to on the way to the site.

Modular building construction has been, and will continue to be, applied to a great variety of structures. Multi-story buildings are candidates for modular construction. Factory-constructed housing lends itself very well to commercial structures. In Geneva, NY a high rise Ramada Inn is being constructed on the shoreline of Seneca Lake, one of the Finger Lakes. See

Figures 2.1 through 2.8, that show the hotel being erected with factory-built room modules. The modules are fabricated in Pennsylvania and shipped to the site ready to be set into place. Apartments, medical centers, fast food restaurants, banks, and condominiums have all been fabricated from modular units. The modular concept puts little constraint on the type of facility.

Figure 2.1
The job site sign with the hotel being erected in the background.

Figure 2.2
Factor-constructed room modules waiting to be lifted into place.

Figure 2.3
A room module being prepared for lifting.

Panelized construction uses factory-built panels that are transported to the site and then assembled into the preplanned unit configuration. Modular and panelized construction can occasionally be blended, and both concepts used to create a structure for housing, or, for other use (offices or retail space).

Manufactured homes are constructed with internal structural support allowing them to be supported by wheels for transport or on piers at a permanent or semipermanent site. Their genesis was the travel trailer of the 1920s. Stephen G. Pappas in *Managing Mobile Home Parks* states, "The invention of the travel trailer was an attempt to alleviate the problems presented by the so-called folding 'tent trailer' which provided little protection from the elements for sleeping, food storage, and camping equipment." Manufactured homes are of single wide or double wide construction. Most single wide homes are twelve or fourteen feet wide by up to seventy or more feet long. Double wide homes are twenty-four or twenty-eight feet wide. The maximum width is controlled by the transportation laws of the state limiting the width of a vehicle on the state's roads. Some states are beginning to allow

Figure 2.4
Inside the room module shown in Figure 2.3. Notice the concrete floor, return air grille, thermostat, and electrical outlet.

sixteen feet wide units. Manufactured homes may be placed and occupied within manufactured home parks, or on property owned or rented by the manufactured home owner. Because it is possible to move the home to another site, lending institutions will loan money for the purchase of a manufactured home.

Most, if not all, manufactured homes are constructed in accordance with the HUD Code as discussed earlier.

Manufactured homes are unique to the U.S. I have traveled in several continents and do not recall seeing a manufactured home park outside North America. The August 20, 1996 issue of *USA Today* cited the growing demand for manufactured housing by retirees and first-time home buyers and the changing image of the industry. The article stated that in 1991 there were 17 million people living in 7 million manufactured homes. By 1995 those numbers had changed to 18 million people in 8 million homes. The average area of a manufactured home increased from 1225 square feet in 1991 to 1355 square feet in 1995.

The August 1995 issue of *Popular Science* reported that Ball State University's Housing Futures Institute in Indiana has studied new ways of assembling manufactured homes. Reportedly in one version, two factory-made sections were stacked, and, with a "bonus" space between, were linked to a third section. A decidedly nonmanufactured home roof line as well as the extra space between the sections was created.

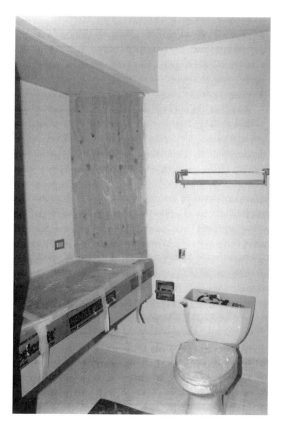

Figure 2.5
The bathroom is ready to go.

Of course, the nonmanufactured home is the home built on the site by carpenters and other tradesmen using 2" × 4" wood members among other construction materials. It is commonly referred to as "stick-built".

The Reputation of Manufactured Home Parks

Historically, manufactured home parks, earlier known as trailer parks, and now becoming manufactured home communities, have had a poor reputation; and, in some areas they still do. Today's manufactured home is far superior to what was constructed 15 or 20 years ago. A smaller unit is a quality piece of affordable housing. Larger units in many cases cannot be considered "affordable housing;" they can become quite expensive.

A reason for the poor reputation is that frequently to obtain the most units per acre of property, the homes are placed very close together and lined

Figure 2.6
Getting ready to lift.

up like dominos; not presenting a very attractive place to live. A strong reason for the poor reputation of manufactured home parks is that all too often the areas were not maintained in a neat condition. Grass was allowed to grow into weeds, junk was permitted to accumulate, etc. There is absolutely no reason why people with low incomes must live in squalid conditions. Neatness does not cost money. Unfortunately, the general public has the NIMBY attitude about manufactured home parks, "Not In My Back Yard!"

How You Can Use This Book

This book is intended to provide guidance to those associated with the development of manufactured home parks and small realty subdivisions. Improved living conditions should be one benefit of understanding the information presented herein. The designers of a development will be able to

Figure 2.7
Lift underway.

create layouts and design the necessary infrastructure in a more timely manner. The developers, realtors, and bankers will have a better awareness of what is required to prepare the necessary documents and what should be contained in those documents. Chapter 1 on the economics of manufactured home park ownership undoubtedly also will interest them. Members of municipal boards with approval jurisdiction may appreciate the treatise regarding the ownership of land and any housing units placed upon the land. The standard details will probably interest designers and contractors.

Major References and Approving Authorities

This book is not intended to replace or countermand any governmental jurisdiction. The development of manufactured home parks and realty subdivisions is highly regulated in the U.S. In many instances each town or parish

Figure 2.8
Moving room into place.

will have its own governing regulations. The authorities, and there will normally be several, in all cases **must** be consulted; and, before construction can begin, the approvals from all interested agencies must be obtained. All persons affiliated with a development are advised to consult their attorneys. On economic questions, guidance from an accountant is strongly suggested.

Major references that should be consulted by any development team will include:

State and local codes, especially health and construction
Water supply and waste treatment must conform to the local authority
Municipal zoning and subdivision ordinances
Local environmental regulations
State highway requirements

Figure 2.9
A modular home.

Figure 2.10
A manufactured home.

Legal Documents

3

Legal Documents

For the purpose of this book, this is my interpretation of the legal documents pertaining to manufactured housing. Please consult an attorney for legal advice.

Deed

A deed is defined as a document sealed as an instrument of conveyance of property. The title to real property is transferred by means of a deed, a copy of which is usually filed with a municipality, typically a county. A deed describes the property being transferred by using the information provided by a land surveyor. A homeowner purchasing a lot in a subdivision will receive a deed stating that title has been conveyed. The owner of a piece of property to be developed as a manufactured home community will retain the deed for that entire parcel of property.

The land occupied by a manufactured home can be simply a parcel of land of any size, although the municipality may have a minimum property size for placement of a home. Some municipalities may not permit any manufactured homes. Usually, a municipality will permit a few homes to be placed on a site. After perhaps three to five, a municipality wants the developer either to create a subdivision or a manufactured home community.

A subdivision is a parcel of land that has been divided by the owner/developer into multiple parcels or lots of land, each of which is legally described by a licensed surveyor and by deed. The municipality will treat each subdivided piece as an individual tax parcel and taxes will be due from the individual owners of each lot. The roads in a subdivision will normally be owned by the municipality. The municipality may, or may not, provide water and sewer service. Although the roads are owned by the municipality and the municipality may furnish water and sewer services, the construction of the roads and of the water and sewers is invariably the responsibility and expense of the developer. The municipality takes ownership after the infrastructure is completed and the construction has been approved by the municipality's engineers. The homes in subdivisions are normally modular or stick-built homes, although I have designed a subdivision developed by a manufactured home community owner/retailer who will only sell lots complete with a manufactured home. On a trip to the western states, I visited several realty subdivisions developed with manufactured homes, some with full basements. The type of structure does not define a subdivision; ownership of the land does.

A manufactured home community is a parcel of property operated to rent space, also referred to as lots, for the occupancy of manufactured homes (typically more than three). The owner of the entire parcel retains ownership of each manufactured home lot. The ownership of the lot is not transferred to the occupants of the manufactured home on the lot. Therefore, deeds are not required for each individual lot in a manufactured home park. And, in real life, the lots are not very well defined; stakes do not normally mark the corners. This is an important difference between a realty subdivision and a manufactured home community. In a realty subdivision, each lot is owned by the occupant and a deed is required for the transfer of ownership of a lot from the subdivision developer to the purchaser of the lot.

A community is complete with water, sewer, and electric utilities extended to each lot for connection to the individual manufactured homes. Natural gas or fuel oil may also be available at each lot. However, environmental spill issues have all but eliminated central fuel oil systems in parks. The overall maintenance of the property is the responsibility of the community owner. Each tenant usually is responsible for mowing the lawn around the home, keeping the immediate area in a neat condition, keeping the home attractive, and similar chores. The municipality within which a manufactured home community is located typically has little responsibility (no snow removal) for services within the park. Water and sewage disposal may be municipal or may be the property of the owner. When community owned, they are frequently referred to as "private public" systems. Electric service may be metered by the utility on one meter to the community owner, or by individual meter to each home owner. The community owner pays taxes to the municipality on the entire community.

Survey

A survey is a determination of the boundaries of a parcel of real property by means of measuring angles and distances using the techniques of geometry

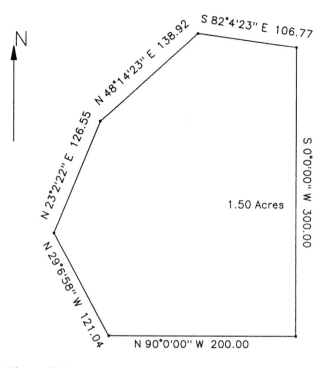

Figure 3.1
A typical survey.

and trigonometry. The angles and distances describe the "metes and bounds" of the property. It may include topographic information that determines the elevations of the land. A land surveyor, one who is trained and licensed by the state in which the work is being performed, surveys the property using appropriate measuring instruments. Traditional measuring instruments have been a transit and a tape measure. The transit allows the surveyor to measure the angles of the property. The surveyor normally transfers the information gathered at the site to a drawing or map, then places stakes or pins at the property corners so that anyone may find the property outlines (see Figure 3.1).

Most local governments will require a field survey of the property, and the staking of the corners of each lot, to transfer title to the property and file the deed with the appropriate authority.

Limited survey information frequently can be obtained using a municipality's tax maps. If only approximate land areas are required, tax maps from the local assessor's office may suffice. Often they are advantageous for preliminary feasibility studies or other such work. I have made extensive use of tax maps when developing manufactured home communities.

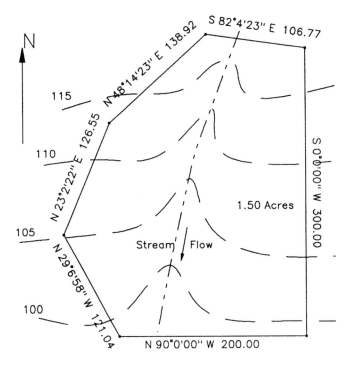

Figure 3.2
A survey with topographical information.

Topographic Survey

As previously mentioned, a survey may include topographic information. Topographic data are obtained by the land surveyor by using a level. The elevations of the various property areas are determined with the level. Then a drawing is prepared that has topographic lines connecting the parts of the property that are of the same elevation (see Figure 3.2).

Computer Technology

Today's computer technology permits topographic information to be determined from an airplane. This method is more economical for large areas than surveyors stomping over the property. To provide the information from the air, a few control points must be field spotted on the ground. Then, by use of equipment that can most simply be described as electronic stereoscopic, the elevations of the land can be determined. This information is then digitized and becomes available for computer aided design software programs to use in the development of surveys. These surveys will show land use with detailed information, including, normally, telephone poles and fire hydrants.

Topographic information is required for the design of manufactured home communities and subdivisions. Homes cannot be sited on steep hills, roads must be constructed with reasonable slopes, sewers must transport fluids downhill.

USGS Maps

The United States Geological Survey (USGS) publishes maps with topographic information and other major ground features. This may include lakes, streams, swamps, roads, major structures, pipelines, trails, etc. In rural areas, individual homes may be indicated. Frequently, much design can be accomplished by use of tax maps and the USGS maps. This is particularly true for manufactured home communities where the individual lots need not be as precisely defined because ownership is not a consideration. A field survey is not always mandatory for a manufactured home park. Also, preliminary planning for a subdivision could be done this way without committing to the cost of a land and topographic survey. Combining USGS maps with tax maps can provide a lot of information at a relatively low cost.

Municipal Zoning

A municipality may create zones for various uses within its municipal area. Typically, there are residential zones, industrial zones, business zones, etc. There may be various types of each. For instance, a zoning ordinance may contain single family, two family, and multiple residence classifications of residential zones. Only the defined type of development is permitted in each zone. The areas of the municipality where manufactured home communities are permitted may be severely restricted. Single, occupant-owned, manufactured homes may be permitted in some residential zones where a manufactured home community would not be permitted.

As with most things in life, there are pros and cons to zoning. Some municipalities are proud they do not have any zoning. Others are rigidly zoned. No zoning allows a property owner much flexibility in the use of the land. Conversely, with zoning, industrial areas are separated from residential areas and the two are not intermixed.

Appendix 2 is reprinted from *Managing Mobile Home Parks* by Stephen G. Pappas and is an excellent discussion of "Laws and Zoning".

Subdivision Regulations

Subdivision regulations are one of several documents that place constraints on developments. As with zoning, depending upon your point of view at any particular time, you may like or dislike a particular subdivision regulation. Subdivision regulations spell out what a developer must provide to a municipality before land can be subdivided. Typically, subdivision plans must be developed to show the municipal authority having jurisdiction (normally a planning board) that the proposed subdivision conforms to the municipality's zoning ordinance, subdivision regulations, health department requirements, environmental concerns, and so on.

Subdivision regulations may include the sizes of residential lots in any of several areas zoned for residential development. Street widths and road construction details may be included in subdivision regulations. Must utilities be buried? Are there buffer areas required? Do recreational park "set asides" need to be deeded to the municipality? Will the storm drainage from the development have impact on any adjacent area? How has the developer protected other properties?

Appendix 3 is a municipalities "checklist" for the items it carefully checks to ensure the developer has addressed the regulations.

Municipal Approvals

Most municipal entities (city, town, county, or other) will have zoning and/or subdivision ordinances restricting development. It would probably be unusual to find a subdivision regulation without a zoning ordinance. One entity may have very restrictive zoning regulations and an adjacent entity may have no zoning whatsoever. A town may have no zoning, but the county in which it is may have restrictions, e.g., health department regulations with respect to water and waste disposal. In most cases, building permits will be required for any construction. A local government, with or without a zoning ordinance in place, will typically allow an individual to construct a home on a parcel of property, and it may allow several homes on a parcel. If the parcel is subdivided, then each home can be on an individual lot. At some point, often three to five homes, a municipality's ordinance will define the project as a "subdivision", a "manufactured home community", or some other such development. And, the fun begins! This book is primarily about such developments.

Obtaining the approval for a subdivision or a manufactured home community from the municipality is an item sometimes overlooked, especially by a person new to the development process. There is, of course, an expense to obtain this approval and sometimes it can be very significant. I have appeared before a town planning board many times over a period of a year or more to obtain approval of a manufactured home community. In another instance, it took close to two years to obtain approval for a subdivision because the municipality wanted an environmental impact study on the property. Each appearance costs money; not only to appear but also to rework the documents to satisfy the changes requested and in the preparation for the appearance. More than one developer has given up, and some have declared bankruptcy, because of the demands of a municipal board.

It helps if the person making the presentation before a municipal board has a bit of a thick skin, especially when presenting manufactured home community proposals. Almost by definition, the public will be against a manufactured home community. Frequently, the public is against any development, no matter how beautiful. Many times I have heard an adjoining land owner rail against a proposed project because it will change the wonderful open space that has existed for all the time the owner has lived there. Once, but only once, the chairman of the board in desperation after many comments by the same owner suggested that if the owner wants to keep the land forever wild, the owner should purchase the land from the developer! An owner of a particular parcel of land has the right to do whatever is desired with the land so long as it conforms to zoning and subdivision regulations and does not have a negative impact on the community as a whole.

In most cases, one or more public hearings will be required before the municipal boards can approve a subdivision or a manufactured home community. This is the public's opportunity to comment upon the proposed project. And, many of the public love to exercise that opportunity. More than once there has been derogatory references to my character. Those comments are generally best ignored. I do not expect a member of a municipal board to treat me in that manner; I expect professional courtesies from the members. The one time when a board member became a bit personal, a private discussion with the municipal attorney after the meeting prevented any future such comments. The municipal attorney usually wields substantial power. The term of service for the attorney typically transcends the terms for the members.

Deeded Subdivision Covenants

It may not be generally realized, but many times there are restrictions established by a subdivision developer on the purchasers of the subdivision lots or homes. The covenants may list:

- The minimum size of a home
- If more than one family can live in a home
- Whether out-buildings are permitted
- If vinyl siding is permitted
- If a doctor's office is permitted

Appendix 4 is a listing of covenants that have been included in various subdivisions. A developer may include just about anything desired in the listing of covenants. Of course, they must all be legal constraints upon the subdivision.

Municipal Home Community Code

In an effort to improve the living conditions in manufactured home communities, some municipalities have created community codes. In addition to permitting manufactured home communities only in specific zones, there may be a community code to control how the community is maintained and the overall appearance of the community. See Appendix 5 for a typical community code.

The first codes developed enforced the code only upon the property owner, the owner of the community, the taxpayer. Those codes did not recognize that, except for eviction, the community owner has little control over a tenant. The code in Appendix 5 includes a method for the code enforcement officer in a municipality to issue a citation to a manufactured home occupant. Now the occupant is responsible for some conditions in the community, especially in the vicinity of the home.

Operator's Manufactured Home Community Rules and Regulations

In addition to any municipal code, most manufactured home communities have their own rules and regulations. These rules and regulations can contain anything the owner wishes to include providing it is reasonable and will be accepted by persons renting lots or homes within the community. Typically, a set of rules will include when payment is due, any designated quiet time, who mows the lawns, who shovels the snow, under what conditions the occupants can be evicted, motor vehicle regulations, how to handle complaints with other tenants, etc. See Appendix 6 for the rules of the Hillside Acres Manufactured Home Community in Ithaca, NY.

Engineering Drawings (See Chapter 13)

In most areas of the country, in addition to the surveyor's map, it is necessary to prepare drawings for any proposed manufactured home development for approval by the municipality, and also by the health department. Such drawings are normally prepared by a licensed professional engineer. As a minimum, the drawings would typically include lot layouts, road configuration, including construction details as required by the municipality, water and sewer details for health department review, lighting systems, etc. If the development is a manufactured home community and not a legal entity as previously discussed, the lots would be shown for the municipality to determine conformance to any code requirement and for the future occupant to have an idea of the property being rented.

Copies of the drawings would normally be maintained by the owner for the future reference of utility systems, and to show prospective tenants the overall plan of the community, as well as the area of the prospective lot being rented.

Construction Documents

Although not part of a manufactured home development, the "construction documents" used to develop and construct a facility are most important. The construction documents begin with some sort of agreement between the owner/developer and the designer. Additionally, the construction documents should include the drawings prepared for the project, specification information, contracts between the owner/developer and the contractor, submittal information on products to be incorporated into the construction, requests for payments, guarantees, etc. The type of document will most likely depend upon who the principals are.

It is well recognized that many times there are no documents for the construction of a development, only a handshake or two. This is the worst imaginable method of operation. Only in a very few instances will there be any significant problem in the development of a project, but, in those instances, it is crucial to have documentation of what was agreed upon and how to resolve any differences. It is well known that there is only one winner in a lawsuit, and it is neither of the parties of the suit.

If the owner/developer is a large regional or national concern, they may have their own documents for construction. If not, there are several standard sets of documents that may be used that have been developed specifically for construction projects. Two are the standard forms prepared by the American

Institute of Architects, and the standard forms prepared by the Engineer's Joint Contract Documents Committee.

Other texts are available and should be consulted for detailed information on construction documents.

Factory-Constructed Home Development Layout:

4

Manufactured Home Communities and Realty Subdivisions

Constraints

Before we spin our wheels and have to do the major portion of the design effort for the project more than once, I strongly urge that we make a list of the constraints on the project. Hang it up in front of the designer(s) so an item is not forgotten that will cause any rework of the layout of the project.

Constraints
Minimum lot size is W' × L'
Minimum road frontage is A'

There are many, and some are not too obvious. It can be very embarrassing and expensive if an obvious, or even not so obvious, constraint does not permit a layout that is substantially developed in the computer or on paper. More than once I have spent many hours on a layout that was looking super only suddenly to realize it was impossible to implement because I had overlooked a constraint.

Constraints come from several directions

Political
Municipal
Legal
Geographical
Financial institution
Market
Owner

Political constraints can be seen and expected and also can be unexpected. A change in the make-up of an approving municipal board can have a devastating impact upon a development.

Many of the municipal constraints are up front in the zoning and subdivision regulations. More on this later. However, what are the requirements of the local health department? Is there municipal water and sewer? When municipal water and sewer are not available, is the health department's minimum lot size larger than required by the zoning/subdivision/home community ordinances of the local municipality?

Geographical constraints include streams, lakes, and probably more commonly, topography. In the Midwest on the plains, topography is seldom of much concern. Elsewhere, in the west and the east, hills and mountains play a very important part in any development. In hilly terrain a grid pattern just will not be feasible. Ironically, neither may a free form layout. A road cannot be laid out to go up a 15% grade in snow country.

Whether or not a lending institution will provide funding for a project is sometimes the go or no go factor for the project. Is the owner in a position to obtain funding?

Are market conditions in the locality suitable for the additional housing proposed?

Unfortunately, the owner's constraint may be the most difficult to determine. This is especially true with an inexperienced owner. Anyone laying out a subdivision or manufactured home community must ask questions of the owner. Then, ask more questions. And, ask them again. The sooner the total project criteria can be established, the easier will be the design and development.

Questions the owner must answer include: What size lots? What configuration, a grid system or free form? What are the owner's esthetic concerns? Are there any recreational facilities over and above what is demanded by the local government? Is there going to be a community center? In a manufactured home community, where is garbage pickup; where are the mail boxes to be? If not demanded by the municipality, what kind of site lighting will there be? Maybe the municipality does *not want* any site lighting; it could have an adverse impact on adjacent properties. With a manufactured home community, will there be double wides or single wides, and in what proportion?

What is the phasing of the project? As municipal boards approve development, the taxes on the property change — the property has potential income compared to open land. Any project should be phased to keep taxes low in comparison to potential sales of the land. Do not obtain approval of a large development of 100 sites which cannot be sold/occupied for six years. Not only are there higher taxes on land approved for a development, but there are roads to build and water and sewer systems to install that have a significant cost. There is no need to have this infrastructure sitting unused while waiting for lots to sell. The industrial business community's "just in time" inventory philosophy cannot normally be achieved, but should be kept in mind. In most instances, local laws require that any property approved for subdividing should have the infrastructure installed within a given period.

It is not necessary for the designer to settle all the questions with the owner during the first several meetings. Some answers are required earlier than others. However, it is most important to determine the criteria early in the planning process. A design for a manufactured home community that includes an attractive entrance with landscaping to the first site, will probably require extensive reworking if the owner decides that mail boxes and garbage pickup is to be near the entrance to reduce traffic within the community.

General

The design of a factory-constructed home development begins with the layout of the project, the arrangement of the streets and lots. Play and recreational areas, if any, must be allocated. During this phase of design, the sewerage system must be considered. Are there public sewers? If so, no space is needed for sewage disposal as the sewer lines will be run in the roadway rights-of-way. If public sewers are not available, calculations are necessary to determine the space required for a sewage disposal system. What are the soil conditions? Soil tests will be needed. It would be most prudent to consult with the local health department. A sewage disposal area must be set aside and should not be used for lots and roadways.

Surveys and topographic information were discussed in Chapter 3. Now is when they will be used. Before lots and roads can be laid out, the perimeter boundary must be known. Also, unless the property is in the middle of very flat country, topographic information is necessary. For the purpose of laying out a manufactured home community, a current field survey is not mandatory information. Remember, property is not being transferred, only rented. A tax map is frequently suitable for this task.

For a realty subdivision, a field land survey and a topographic map will be mandatory. A municipality will frequently want topographic data in 2-ft increments. The land survey must be suitable for the surveyor to describe each lot for the purposes of a deed and to set stakes at the lot corners.

The United States Geological Survey (USGS) publishes maps of the U.S. that show the land with roads, buildings, water areas, swamps, wooded areas, pipelines, municipal boundaries, and, for the purpose of this discussion, topographic information. They are usually drawn to a scale of 1" = 2000'.

Figure 4.1
Part of a United States Geological
Survey (USGS) map: 1″ = 2,000′.

The maps measure about 17 in. by 23 in., cover about 55 square miles, and
are an extremely valuable resource. They are available from most large sta-
tioners. Although the accuracy of the topographic information on the maps
is not nearly as good as a field or aerial survey of the specific site, the
information is normally more than adequate for use in the design of a
manufactured home community and for much preliminary information for
a subdivision.

An inexpensive method of obtaining reasonably accurate survey and
topographic data is to combine a municipal tax map with a USGS map. This
scheme can be used for preliminary analyses of subdivision layouts or for
many final drawings for manufactured home communities. If a USGS map
is enlarged 40 times, the map scale becomes 1″ = 50′, a very acceptable scale
for the design of a community. A tax map from the local municipality with
a scale of 1″ = 400′ can also be enlarged to 1″ = 50′. Combining the two maps
results in the basic information needed to begin design of a development.
The combined maps can be traced onto drafting paper, or, with the current
state of the art, can be scanned into a computer and then used for design
with a computer aided design (CAD) program.

Figure 4.1 is a portion of a USGS map with scale of 1″ = 2000′. Figure 4.2
is the topographic map enlarged, via a photocopy machine, in this case five
times, to a scale of 1″ = 400′ to match the scale the municipality uses for tax
maps. Figure 4.3 is a copy of a portion of the tax map showing the owner's
property. Figure 4.4 is a copy of a sandwich of a transparency of Figure 4.2
and Figure 4.3. Figure 4.5 is a print from a CAD program of the scanned
image from Figure 4.4. And, we finally have a usable document; Figure 4.6
is the improved computer image of the property with topo and property line.
Instead of scanning the document for the CAD program, a digitizer could
also be used to accomplish the same result. Actually, with the digitizer, you
would go directly from the sandwich, Figure 4.4, to the final drawing,
Figure 4.6.

Another resource for the basic property information is aerial maps maintained by some municipalities. Figure 4.7 is a copy of an aerial map for the
same property. The information on the aerial map will probably be more
current than the USGS map, will show better wooded vs. open field areas,

Figure 4.2
Part of a USGS map enlarged to 1″ = 400′. (Reproduction @ 75%)

and can confirm further the property features that should be considered for design.

Remember, as discussed in Chapter 3, aerial surveys are excellent for design purposes. There may already be aerial information available from the local utility or municipality. The owner may be able to purchase the digitized information for the parcel of land being developed. This is, by far, the preferred source of data for professionals developing drawings with CAD software.

Layout of the lots can now begin.

Typical Considerations

The layout of a realty subdivision and of a manufactured home community are similar. The major differences will normally be in the size of the lots and

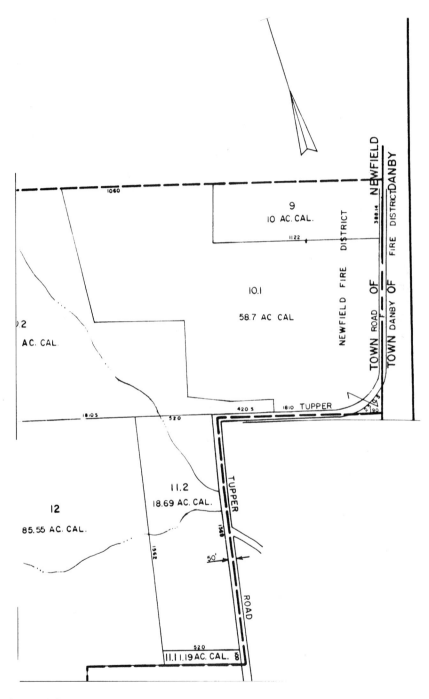

Figure 4.3
Part of a tax map: 1″ = 400′. (Reproduction @ 63%)

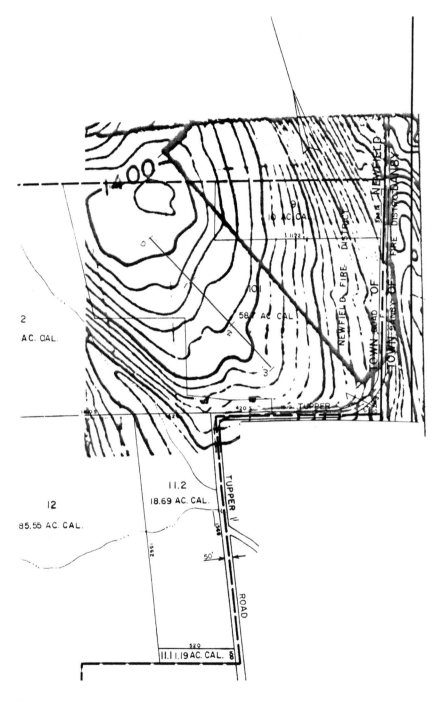

Figure 4.4
Copy of a "sandwich" of Figures 4.2 and 4.3.

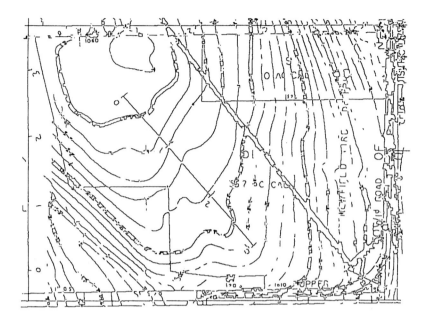

Figure 4.5
Computer-generated print of scanned image of Figure 4.4.

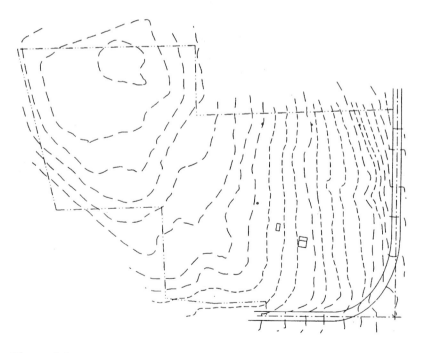

Figure 4.6
Enhanced image of Figure 4.5; ready to start design.

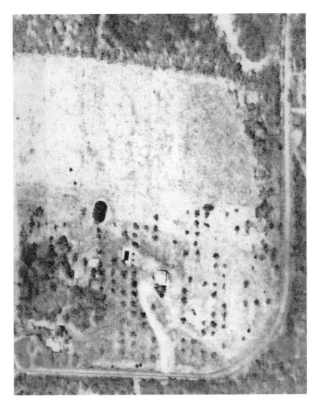

Figure 4.7
Aerial photograph of the property.

the municipal constraints placed upon the development by any zoning and subdivision regulations of that municipality.

Typically, the major constraint placed upon a manufactured home community is the area in the municipality where any zoning ordinance would permit a manufactured home community to exist. Beyond this, many municipalities will be found to have little regulation of how a manufactured home community is laid out. Others will have details for a manufactured home community that are as similar in extent and complexity as for a realty subdivision (again, see Appendix 5).

Considerations that you may find in municipal ordinances regulating manufactured home developments include:

- Setback distances from adjoining land owned by others to a home
- Setback distances from a public highway to a home
- Setback distances from a development road to a home
- Separation distances between two homes

- Road widths, possibly including widths of driving lanes and shoulders; also, the finished driving lane surface may be specified
- Road corner radii and cul-de-sac radii (in a manufactured home community, it may seem that the various radii are only of concern to the community owner. However, emergency vehicles, delivery vehicles, and the like, must be able to traverse a manufactured home community as well as a realty subdivision.)
- Recreational areas
- Amenities for each lot, i.e., parking spaces, storage areas, outside lighting

All these considerations, and most likely many others, must be considered when designing a development. The question that the designer must answer is, who is stating the criteria? The owner, the municipality, the health department, or the designer?

The municipality may ask for comment/agreement from the local school district, fire district, a separate water purveyor, neighboring property owners, and others.

Additional constraints upon the layout of a development may be placed by the local governing department of health. If a private sewage disposal system is to be constructed, there may be minimum separation distances between homes, septic tanks, septic fields, wells, water piping, sewers, etc.

Development Design

Now we can be creative.

The design for the arrangement of the lots for a manufactured home community or for a subdivision is the fun part of a development project. The possibilities for lot configurations are almost endless. This is where the designer can go berserk and create an attractive and enjoyable development that appeals to a variety of people. Or a dull, every lot the same size and shape, development can be the result of a lazy designer. A word of caution for anyone looking at the plan of a development. The owner/developer may have insisted upon a specific lot arrangement. The design professional may not have had much to say about the final configuration. I have walked away from such an owner.

A development is not just an arrangement of lot boundaries with water, sewer, telephone, and taxes. There must be a certain ambience to the lot arrangement which can contribute to the pride of ownership of a piece of property in the development. Everyone wishes to be different to a degree. Purchasers would be somewhat turned off if every home in a development had the same front elevation and was painted yellow. The same is true of lot configuration. We like to be a bit different; but most of us do not wish to be extremely so.

Naturally, every owner/developer wishes to have thirty lots for every acre! Even those owners who are realistic (truthfully, most are) forget that roads, required open space for recreational parks, and such, reduce the actual useable area of a piece of property. I have read articles that try to prove that developments with curved roads with lots of different configurations can actually contain as many lots as the one size and shape rectangular lot development. I don't buy it. It is impossible for me to accept that a development with curved streets and different sized and shaped lots can yield the same number of lots, with a minimum size to satisfy zoning, as the rectangular lot configuration. At the same time, I believe a development with varying lot shapes can appeal to a much wider audience and provide a much happier environment for the property owners.

Lots with road frontage on three sides should be avoided. If, for some reason, a three-sided lot is absolutely necessary, make it as large as possible. Lots with roads on the front and back should likewise be avoided. Typically, it is desirable to have corner lots a bit larger than the interior lots. Roads should intersect at right angles; when less than perpendicular it is easy for a driver to see traffic approaching from one direction but difficult from the other. Also, it is difficult to turn into a road when the turn is greater than ninety degrees. Curves should be gentle to make for easy traffic flow and good sightlines; radii of 120 feet or so are suitable. Sightlines over hills and around natural features must be considered carefully.

Figures 4.8 to 4.11 demonstrate comparisons of manufactured home lots, modular subdivision lots in a rectangular pattern and in nonrectangular arrangements. Figure 4.8 shows 108 sites for manufactured homes in a 39.6 acre parcel, which is about 1750 ft by 1000 ft. For this comparison, level property was assumed and the topography was not a consideration. Whenever land is being developed, the zoning ordinances and subdivision regulations of the municipality must be considered. Typically, in my area of practice, manufactured home regulations specify a minimum lot size of 5000 square feet. Here we have a case of the market dictating a higher standard than the ordinance. Lots of 5000 square feet are no longer popular or practical with today's larger and better manufactured homes. The lot size used for this comparison is 65 ft by 110 ft, 7150 square ft.

Figure 4.9 demonstrates that only 58 realty subdivision sites can be offered on the same parcel. Again, the laws of the municipality must be considered. Here, a zoning regulation of a minimum lot size of 25,000 square feet with minimum frontage of 100 ft and minimum depth of 200 ft was used for the design.

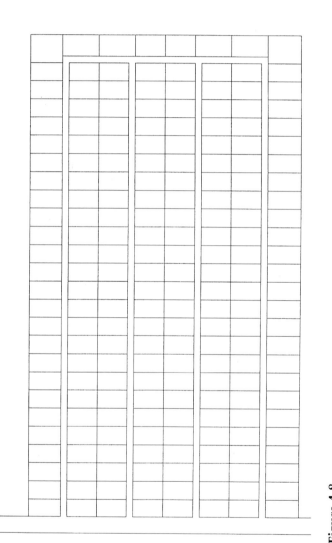

Figure 4.8
An arrangement of manufactured home lots in a 39.6 acre parcel with one hundred eight 65' × 110' sites.

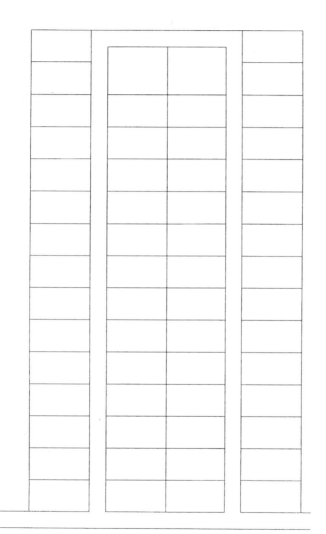

Figure 4.9
An arrangement of realty subdivision lots in the same 39.6 acre parcel with fifty-eight 115′ × 220′ sites, 25,300 square feet each to satisfy zoning requirements of a minimum of 25,000 square feet.

Both examples are based upon a pure rectangular arrangement of the lots, which I believe does not improve the esthetics of the development. However, the examples do point out a factor for affordable housing. Affordable housing is not solely a function of the cost of the home. It is also a function of the cost of the land, the cost of infrastructure, the cost of interest, etc. The infrastructure for a parcel with smaller lots has a lower cost per lot when there are more lots per foot of road frontage. When a 1000 ft long sewer can serve 15 lots instead of 10 lots, the cost of the sewer, per lot, is reduced. Typically, the size of the sewer does not normally change when a few more pieces of property are added. Eight-inch sewers are the normal smallest municipal sewer installed because of sewer slopes and the ability to clean a sewer. The smallest water service is normally 6 in. to provide adequate water for fire protection.

Figures 4.10 and 4.11 demonstrate realty subdivision arrangements with a more pleasing road configuration which leads to variously shaped lots. These road layouts are not particularly liked by the highway departments, especially in snow country. It is much easier to clear snow from long straight roads than from curved roads and particularly those with dead ends. But, I believe, the quality of life is much improved with nonrectangular lots, be it for a manufactured home arrangement or for a modular subdivision.

A method of reducing housing costs and increasing the affordability is to develop "cluster" housing. And this can be applicable, to a degree, with manufactured home communities and with realty subdivisions. Generally, cluster housing places the allowable number of lots for a piece of property within a portion of the property and leaves the remainder as open space. This results in smaller individual lots but reduces the cost of the infrastructure. It also creates open space that can be used by the occupants for recreational purposes.

Figure 4.10
A nonrectangular arrangement of subdivision lots in the same 39.6 acres, showing 55 lots with a minimum of 25,000 square feet.

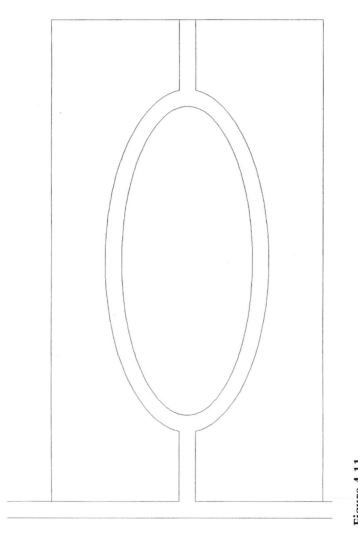

Figure 4.11
Another possible roadway configuration in the 39.6 acres.

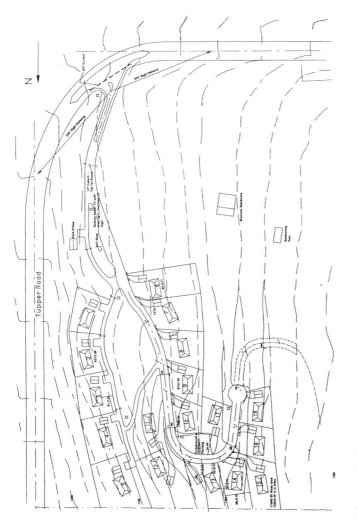

Figure 4.12

An 18-unit double wide manufactured home community on a hillside developed from Figure 4.6. [Reproduction @ 65%. Format represented is important; specific text is not.]

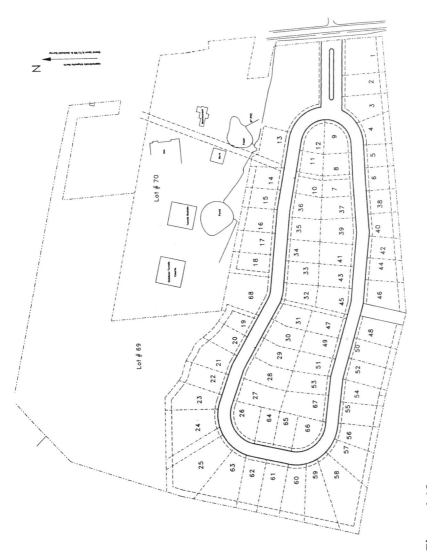

Figure 4.13
A 70 lot subdivision in a cluster arrangement. [Reproduction @ 62%. Format represented is important; specific text is not.]

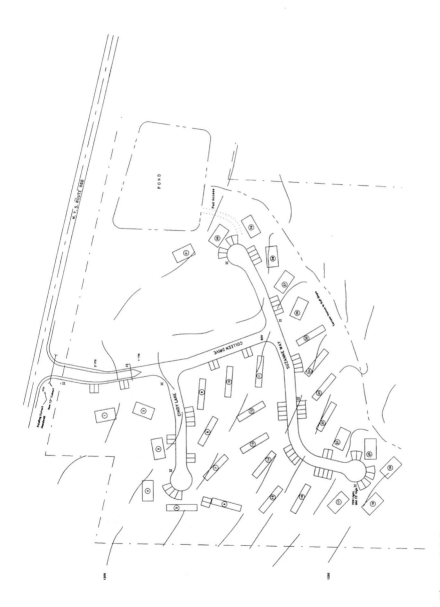

Figure 4.14

A 31-unit manufactured home community, single and double wide sites. [Reproduction @ 62%. Format represented is important; specific text is not.]

Figure 4.15
A portion of the community shown in Figure 4.14.

Figure 4.16
A digitized aerial survey of a portion of a manufactured home community.

Figure 4.17
A portion of the community in Figure 4.16.

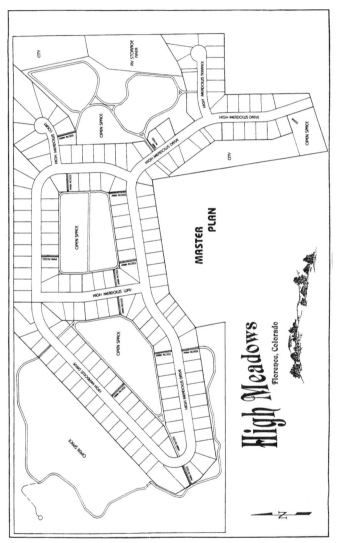

Figure 4.18
The master plan for a 169 home realty subdivision on 92 acres. All homes are to be modular or manufactured.

Figure 4.19
High Meadows Terrace in the subdivision shown in Figure 4.18.

Figure 4.20
A manufactured home in the subdivision shown in Figure 4.18.

Figure 4.21
Manufactured homes on a hillside.

Figure 4.22
Manufactured homes in the woods.

Figure 4.23
Manufactured homes in Ft. Myers, FL on a canal.

Figure 4.24
Manufactured homes in Naples, FL.

Figure 4.25
A modular unit, partially assembled.

Figure 4.26
The basement of the modular unit shown in Figure 4.25.

Support of Factory-Constructed Housing

5

General

More than 50 years ago, factory-built housing included Sears, Roebuck & Co. panelized homes shipped from a factory and trailers with wheels, as cited in Chapter 2, which were developed to replace folding tent trailers. Sears, Roebuck panelized homes were supported on slabs, crawl spaces, or basements, much as today's modulars and panelized units are. The trailers, and that is exactly what they were at that time, were supported on wheels. You could pull into a park and stay the night. The next day you were off to someplace else. Today, the equivalent is known as a recreational vehicle. Today's manufactured home, that has evolved from those trailers, is seldom supported on wheels after delivery to the home site.

Manufactured Homes

The standard support detail for manufactured homes for many years has been a short column of concrete blocks about 2 ft or so high set on undisturbed soil. Wood wedges were used to do the final leveling of a home. Appendix 8, Standard Drawing Details, contains "Manufactured Home Support Sketch" that has been on my drawings for a long time. Figure 5.1 shows a typical support pier with the solid concrete block as the base. Similar details are furnished by the home manufacturers. The National Conference of States on Building Codes and Standards (NCS BCS) A225.1, an ANSI-approved document, contains similar details. Figures 5.2 and 5.3 are similar block supports with concrete and with a railroad tie as the base. Each community owner has a preference. The detail in the Appendix indicates a two block wide support vs. the one block shown in the photos. Higher piers need a

Figure 5.1
A concrete block pier manufactured home support.

Figure 5.2
A concrete block pier set on concrete.

wider base. Many manufactured homes will continue to be supported in this fashion for years to come.

Manufactured homes, especially single wides, are susceptible to being blown over in high winds and must be tied down. The requirements for support of manufactured homes must always consider wind conditions. The

Figure 5.3
A concrete block pier set on a railroad tie.

NCS BCS standards have different requirements along the coasts where hurricanes are likely. As shown in the appendix, my sketch requires a tie down on a diagonal to secure the home properly. This requirement is no different from that required for stick-built homes that are tied down by the bottom plate being bolted to the concrete footing/foundation.

Steel supports are now on the market and are replacing the concrete block system. The supports are basically small four-legged stands with a leveling screw on top to level the home. The top can be fitted with a clamp so when the stand is secured properly to a concrete base, the stand can "hold down" and support the home. I believe a diagonal support is just as economical and a more secure way to "hold down" the home in high winds. Figure 5.4 shows the beginning of a realty subdivision for manufactured housing. The concrete "footings" have been poured for one home. This is in a part of the country that does not have deep frost, which permits shallow footings. The footings are wide to allow support of steel support stands and concrete block skirting.

Figure 5.5 is a close up of a steel support stand for a home in a different place but with similar footings. Figure 5.6 is another photograph of steel support stands and the concrete block skirting from under the home. Figure 5.7 is a photograph of a manufactured home with concrete block skirting. The concrete block skirting, while attractive and most durable, is economical only where deep footings are not required. When the footing needs to be only about 1 ft or so deep, the continuous footing is reasonable.

Figure 5.4
Footings for the first home in a subdivision.

Figure 5.5
Steel support stand.

In colder climates requiring 4 ft (or deeper) footings, a footing to support a concrete skirting would be extremely costly.

Setting manufactured homes on basements was unheard of years ago. Although not a common occurrence today, it is being done. Access to the basement is occasionally through a space left inside the home but more

Figure 5.6
Steel support stand with concrete block skirting.

Figure 5.7
Manufactured home with concrete block skirting.

Figure 5.8
Manufactured home with a basement.

commonly to a small extension added to the home. Figure 5.8 shows a home recently set on a basement. Figure 5.9 is a photo of the unfinished end wall showing the steel cross beams on the concrete wall that support the manufactured home's steel beams.

Modular Homes

The support system for modular housing is a bit simpler, although probably more expensive in most instances than for manufactured housing. The support system for modular housing is basically the same as for conventional "stick-built" homes. The perimeter of the home and any major walls need support from below the frost line up to the base of the wall. A poured concrete or concrete block perimeter plus interior support columns as needed for major walls is the usual approach.

Figure 5.10 shows a poured concrete perimeter with a cross concrete block support system. A row of wood supports goes the length of the center of the home. The photo shows a crawl space that could just as well have been a basement.

The crawl space for a different modular home is shown in Figure 5.11 and the exterior of the same building with an asphalt style coating is shown in Figure 5.12.

Figure 5.9
Basement steel supporting a manufactured home.

Figure 5.10
Support for a modular home.

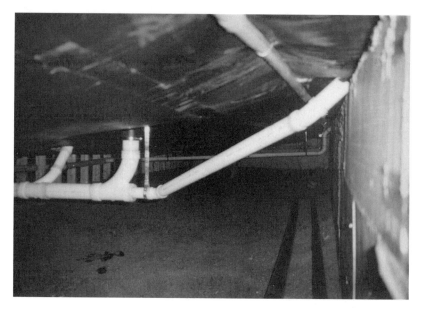

Figure 5.11
Underneath a modular home.

Figure 5.12
Modular wall foundation.

Groundwater Resources

6

Brayton P. Foster

General

When municipal water is unavailable, surface waters or groundwater become the source of needed water. Because surface water bodies are not isolated from potential sources of contamination (requiring construction of treatment systems) groundwater becomes the preferred choice. Groundwater quality and quantity vary widely, requiring exploration and evaluation of this subsurface resource.

Groundwater distribution is controlled by the geology of the subsurface unconsolidated (non-bedrock) sediments and the underlying consolidated bedrock. The volume of groundwater available for use is a function of the porosity and permeability of the subsurface materials. In unconsolidated sediments, granular sands and gravels can contain groundwater. In bedrock, porosity is classified as primary if fluids can be contained in the matrix of the rock (usually sandstone), or secondary if the rock is impervious and fluids can be contained in naturally occurring fractures. Various combinations of primary and secondary porosity can exist and any porosity needs to be interconnected for permeability to be effective. Even when porosity and permeability exist, water can be recovered only from the saturated zone below the watertable. The level of the watertable is a function of several controlling elements, including seasonal precipitation, recharge pattern, outlet elevation, and lateral extent. Man-made changes to the watertable can occur by pumping, changing the outlet elevation by damming or draining, and dewatering for mining or construction.

Unlike the petroleum industry where a hydrocarbon reservoir has a finite volume that when pumped depletes the reservoir, groundwater is a "dynamic" system that is driven by precipitation percolating into the ground to replenish water flowing to lower surface discharge elevations (springs) or water wells.

Groundwater can be captured for use when it flows to the surface, creating a spring. If a spring can be isolated from potential surface contamination by construction of an enclosure, it can be a satisfactory water source if the volume is sufficient. Although some springs produce at a constant rate year-round, many fluctuate seasonally and are at their lowest rates (usually summer) when water demand, especially for watering plants or animals, is usually highest. A variation of the spring is the infiltration gallery. That is a buried trench usually containing a perforated pipe, gravel backfilled, and covered with impermeable soil. If surface contamination can be avoided those structures can connect with shallow groundwater, usually along streams that can recharge adjoining gravel layers connecting with the infiltration gallery trench.

When no surface water bodies or springs are available, a well for water supply can be developed, depending on the depth to groundwater and the subsurface materials. Wells can be dug by hand, or manually driven using a perforated pipe in the lower section of the drive pipe (wellpoint). A common method is to drive the wellpoint with a sledgehammer. Today, wells are most frequently drilled by a drilling contractor. Residential and small community wells can be constructed by any of those means.

You need an understanding of the geology of the specific site before drilling wells in search of water. An initial geologic study to find well drilling sites should be based on air photos, soils maps, local and regional geological maps, groundwater or aquifer maps. An experienced air photo interpreter can readily identify glacial landforms if present, and relate land use and geologic outcrops to topography. In glaciated areas, soils maps are important in distinguishing granular sand and gravel areas from glacial till or lacustrine (lake bed) areas. The gravels and sands can contain water and/or be excellent recharge areas, whereas tills and fine grained (frequently silt or clay) lacustrine deposits have much slower water recovery and percolation rates. Previous mapping, including local and regional geologic maps, watertable or aquifer maps, wellhead protection district maps, and local site specific mapping using available well records all provide data that can be interpreted to aid in a groundwater investigation. Once an initial site has been selected and a well has been drilled, the geologic understanding of the site will be either confirmed or modified to incorporate the data from the new well.

Once the presence of groundwater is established for a specific site it becomes necessary to pump test the well to determine the maximum recovery rate, and more important, the sustained recovery rate over a period of at least one day, and preferably two to three days. Depending on the situation it may be necessary, especially for larger projects, to have at least one and usually more offsetting wells to monitor for drawdown of the water level during pumping. If the site has no preexisting wells it may be advantageous to drill those monitoring wells. While small diameter wells can be used to monitor pump testing, my experience finds that normal 6" diameter water wells are not more expensive yet provide the added flexibility of being able to be pump tested if that becomes necessary now or in the future.

If pump testing methods are not dictated by local regulations, an initial pump test probably will be a step rate test to evaluate drawdown at successively higher pump rates. If equilibrium conditions of a fixed pump rate and stable water levels are reached, the longer the well is pumped at this rate the more confident the tester is of the formation's ability to sustain its rate of water production. Pump testing (and interpretation of pump tests) is a complicated science that does not need to be detailed here. If you desire detailed information on pump testing and interpretation, *Groundwater and Wells*, by Fletcher G. Driscoll, published by Johnson Filtration Systems, Inc. would be a suitable reference. Once initial pump testing is completed it will be necessary to compare the projected water use for the proposed project so that you can make a judgment about the need for additional wells. High demand periods during the day can be dealt with by storage tank capacity, but the well or wells must be able to deliver comfortably the daily water requirement in 24 hours, and frequently in 12 hours as required by regulations requiring surplus capacity. Typically, for larger projects, the daily water requirements should be met assuming the highest yielding well out of service.

Well Construction

What type of drilling operation and ultimately which type of drilling contractor to hire will be debated forever. The basic choice in drilling methods is between cable-tool or rotary drilling. Each method has advantages and disadvantages depending on the situation, and should be selected based on the needs of the specific site.

Cable-tool drilling works better in unconsolidated sediments where advancing the casing during drilling is required, and in sensitive areas where a minimum of disturbance to the water-bearing formation is desirable. A cable-tool rig repeatedly drops a weighted cutting bit into the hole being drilled. It thereby breaks down the sediments to create the wellbore. (See Figures 6.1 and 6.2.) Additionally, an empty cylinder with a foot valve can be lowered into the well to bail out the accumulated water and sediments. (See Figures 6.3 and 6.4). The generally lower cost per hour of operation can make cable tool rigs desirable for detailed sampling, testing, and screening operations, particularly in fine-grained thin layered unconsolidated sediments.

Rotary drilling involves turning the drill bit and circulating air or fluid (water or various drilling "muds") to remove cuttings from the well. Air drilling tends to be faster, avoids the expense of the "mud", and if sufficient air is used to keep the hole clean, particularly in bedrock, is effective in finding

Figure 6.1
A cable drilling rig.

water. Compressing the large volumes of air needed in air drilling requires expensive compressors, and consumes fuel. Drilling with water or "mud" is necessary when the borehole must be kept full of fluid to provide hydrostatic back pressure on loose sloughing material or artesian flows before casing is installed. In the oil industry elaborate "mud" systems are often necessary to deal with hole conditions, and portions of that technology can be used in water drilling when needed.

There is no substitute for local experience if the equipment is adequate for the job and the driller is receptive to trying something new. The selection of the circulating medium should be based on the anticipated depth to bedrock, the stability of the unconsolidated sediments above bedrock, and the characteristics of the water-bearing formations.

Water wells typically have 6" casing set to bedrock, or the top of the water-bearing interval, if bedrock is not reached. This casing is primarily to hold back unconsolidated sediments to maintain the stability of the wellbore. Casing can be driven into place or, if grouting is required, lowered to its final

Figure 6.2
A carbide bit on the end of the cable.

position. Grout, an expanding cement-bentonite slurry, is then pumped down a tremie (grouting) tube outside the 5" casing, or it can be displaced with water down the inside of the casing and back up around the outside. After the 6" casing is set in place drilling can be continued until water is reached. In some situations (such as mud rotary drilling) the casing is not lowered into the well until after drilling is completed. Appendix 8 contains a sketch of the cross section of a relatively uncomplicated bedrock well.

The relatively simple 6" well shown in Appendix 8 can be modified as necessary to deal with conditions encountered (or expected to be encountered) during drilling. If deep depths, large capacity pumps, or severe conditions are anticipated, start with larger diameter hole and casing sizes. If subsurface conditions are unknown for a specific site, a 6" well to obtain information should be the first step, even if this 6" well later becomes only an observation well, before attempting larger diameter drilling.

Looking first at wells completed in unconsolidated sediments (including thick glacial outwash deposits and deep valley fill deposits), the main objective is to advance the casing to the top of a water-bearing layer. Then water can be extracted from the formation without production of overlying fine grained sand, silt, or clay materials. Those tend to accumulate around pumps in the wellbore and in storage tanks. If that drilling is done with an air rotary rig or a cable-tool drilling rig (using the method of drilling a few feet below the casing and then driving the casing down to the depth drilled), the friction on the casing can often limit the length of the casing to between 150' and 300', depending on the materials encountered. If greater depths are

Figure 6.3
The foot valve on the end of the bail.

anticipated, a larger diameter casing (at least 2" larger) can be used for the
initial 100' or so of hole, significantly reducing friction on the casing in the
portion of the well inside the initial casing. Three successively smaller casings
were used successfully to advance the shallow portion of a natural gas well
in more than 500' of thick lacustrine deposits at the south end of a Central
New York Finger Lake. In this specific well, water zones were cased-off as the
water was not wanted in the gas well, but could be recovered later by perfo-
rating the casings after gas production had ceased.

Rotary drilling with water or "mud" circulation can be an alternative in
deep sediments so drilling can be completed before a single casing is installed.
This method also allows the installation of a screen on the bottom of the
casing if needed. In this situation the annular space between the diameter of
the hole drilled and the outside diameter of the casing may have to be
cemented or grouted to prevent fluid movement outside the casing, especially
if artesian pressures are encountered.

Figure 6.4
Bailing a well during drilling.

Casing diameter also generally needs to be increased if the proposed pump capacity exceeds 120 gallons per minute, as pumps in this capacity range usually exceed 6" in diameter. A larger casing size also increases the options available to deal with large boulders possibly encountered during drilling.

The volume of water stored in a 6" well is approximately 1.47 gallons per foot, whereas the capacity of an 8" well is 2.68 gallons per foot, and of a 10" well 4.1 gallons per foot. In residential situations where surface storage is minimal, drilling a larger diameter well just for storage may be economically practical. In commercial wells where larger recovery rates are needed and surface storage usually exists in conjunction with treatment requirements, the added cost of the larger wellbore and associated casing just for the storage is not justified.

If, during drilling, the casing is advanced too deep, shutting off a water-beating layer, and the casing cannot be pulled back because of rig size or casing strength, the casing can be perforated (to allow water entry) by a drilling tool that punches slots into the pipe. Shaped explosives, called "perfs" or shots, run into a well on a wireline by a specialty contractor, also can be used as they are in the petroleum industry, but are generally more expensive.

Some water-bearing sands and gravel are sufficiently sorted to allow water to flow freely to the well, but many water-bearing sands or gravels also contain fine material that needs to be filtered out as the well is developed (cleaned) for maximum production. If a substantial portion of the water-bearing sand or gravel is coarse grained, extended pumping can remove enough of the fine materials to allow the coarse granular material to accumulate at the wellbore, thus creating a natural gravel pack. The dumping of pea gravel (No. 2 stone) into the well before pumping can create an inexpensive gravel filter to prevent fine sediments from entering the wellbore. Such gravel fill may need to extend 5' or more up into the well casing to create a good filter. Commercial well screens made of plastic or stainless steel are also available for installation in wells, to hold the wellbore open while holding back fine sediments. Screens are usually the preferred method when the maximum possible recovery rate and long-term well integrity are required. Screens can be attached by threads, glue, or welding on the bottom or in the middle of a casing string, or telescoped out the bottom of a casing. The selection of the proper screen size requires a sieve analysis of the particles present, so the screen-slot size holds back approximately 40% of the formation, allowing the finer materials to be flushed out by pumping. Slots cut into steel casing with an acetylene torch also can be effective in the right situation.

When casing reaches bedrock without finding water in the overlying unconsolidated sediments, the bedrock will need to be drilled in search of water. Casing is usually driven or set in what the driller considers to be firm unweathered bedrock. Failure to seat the casing into solid bedrock can be the source of leaks that can allow contamination to enter a well, or loss of water.

If the water-bearing zone is a specific geologic rock formation, drilling is continued into bedrock until that formation is penetrated. The recovery rate is then pump tested, and if sufficient, the well is ready for use as soon as it is clear of the cuttings and cloudiness created by drilling. If water is produced from natural fractures, the depths to those fractures may or may not be consistent between wells, depending on the fracture characteristics of the rock. Depths to potable water vary widely, with some areas being restricted to as little as 35' depths, while other areas of deep water zones can be 1000' or more deep. If fractures produce water and pass the pump test, the well is ready for use as soon as the water is clear.

Pump testing can be done by several methods, such as air lift pumping, bailing, swabbing, or pumping depending on the construction of the well and the equipment available. In any pump or production test you need to record initial conditions, and frequently, production rate, the water level in the well vs. time, and well conditions after the test stops. Add whether or not the well recovered to its original level. It is also important to recognize that an initial recovery rate measured as soon as the well has been emptied of stored water can be very different from a stabilized pump rate determined at the end of an extended period pump test. Thus, it is important to recognize how a pump or recovery rate (such as a driller's rate) was arrived at, when comparing wells.

Well Quality

To this point the emphasis in this chapter has been on finding water and well construction, but the other half of the water equation is water quality; can you use it or drink it? Water quality can be divided into three parts: suspended particles, bacterial contamination, and dissolved minerals. Any of those quality problems can prevent water from being potable without treatment, or the well is useless.

Water quality testing begins as soon as the drill cuttings have been removed from the well. The source of particle contamination needs to be identified to determine if it can be eliminated from the well. If not, the particle contamination will need either filtering or a large settling tank. Colloidal or suspended clay that will not settle requires chemical flocculation and filtration, which quickly becomes expensive if any appreciable volume of water needs to be treated.

Bacterial contamination is an immediate health department concern. Depending upon the specific types present, bacterial contamination may or may not be treated adequately by a simple chlorine system, usually required in any municipal system.

Dissolved contamination can range from calcium that causes water hardness, to salt, iron, sulfur, nitrates, and heavy metals. There are U.S. Environmental Protection Agency (EPA) and state drinking water standards for all soluble contaminates and prescribed test procedures for each. If any contaminates are over threshold levels, treatment costs need to be added to your operating budget. Included in those treatment costs is the need for additional water as some treatment systems require water for backlashing that is then discharged.

Pesticide and organic contamination are not common, but need to be tested for in municipal systems. The cost of water quality testing can be great if several wells need to be tested regularly, suggesting one or two good wells are the goal for system efficiency and economy.

If poor quality water is discovered in an initial well, additional explor-
atory drilling should be considered to attempt to find different water that
will require less treatment.

When more than one level of water entry is present in a well, all the water
can be of the same quality, or water from each level can be of different quality.
Technology exists to isolate zones of water entry with at least a few feet of
separation in the wellbore. If a well went too deep, encountering poor quality
water (salty or high sulfur content are common examples) the wellbore can
be plugged back with a mechanical plug or cement to a shallower level.
Elimination or isolation of a shallow water (sediment or bacteria are common
shallow contaminants) zone or casing leak can be accomplished with packers
acting as a liner (many types exist) on smaller diameter pipe. If undesirable
shallow water is known to exist the initial casing can be extended below the
level of that water by planning ahead and being able to drill a larger diameter
hole even in bedrock if necessary.

Recovery Rate

If the recovery rate of a completed well is insufficient, several methods to
enhance the recovery rate exist. These methods are treatments with explosives, dry ice, nitrogen, and hydraulic fracturing. All of them use pressure to
open existing channelways or create new ones. The application of pressure
in excess of 100 psi is usually restricted to bedrock wells. Take care not to
raise the casing like a hydraulic cylinder when pressure is used in treating a
well. Hydraulic fracturing (petroleum industry technology) has the widest
range of applications of the various treatment choices because it is basically
pumping water at whatever pressures are necessary. Explosives and dry ice
treatments will work only if the fractures needed to be opened are not more
than a few feet from the wellbore. Surging (involves using a drill rig to raise
and lower a tight fitting tool in the wellbore in a plunger-like action) is also
a stimulation method that can best be done just after the completion of
drilling, while the drill rig is still over the well.

Acid treatments to remove mineral encrustations (similar to lime in a
tea kettle) are also effective, particularly in cleaning screens.

Maintenance

Some wells will last "forever" with virtually no maintenance, while others
require regular maintenance or cleanouts to remove iron fixing bacteria, scale,
or sediment. The regular shocking with chlorine (as frequently as once a
month) of a well is required when iron bacteria are fouling the pump intake.

The distribution and severity of iron bacteria vary with water chemistry and appear to be worse in wells that are routinely pumped down to the pump intake level. That suggests that the introduction of oxygen invigorates the bacteria.

Assuming the water in your well is potable, the presence of leaks in the piping system can be the major concern of health department regulators. Leaks can arise from sub-standard pipe, poor vents, careless backfilling of trenches, corrosion, accidents, etc. Most of them can be prevented. Equally important is the need for valves, gauges, and faucets that allow individual components of a water system to be evaluated when a problem arises.

Another area that cannot be overemphasized is record keeping. Details of initial construction; water levels before, during, and after pump testing; and production rates must be preserved. Weekly and preferably daily production records, water levels, and energy use can be used for comparison when a problem is suspected in the future. A sudden surge in use can indicate a leak.

This discussion of water wells will not make you an instant expert. Rather it will show that a world of technology exists for use in water well construction, and once a well is drilled the owner is not stuck with whatever results the first well produced. You need the assistance of people knowledgeable in geology, well construction, well maintenance, and buried plumbing, particularly if they are employed during planning and initial exploration stages. A water well is more than a hole in the ground.

Well Log

Figure 6.5 is a sample well log that the well drilling contractor should complete for the owner's records. The local health department will also want a copy. Many well drillers and health departments have their own versions of well logs, and most will document the well. Other documents may include yield logs and drawdown logs.

WELL   LOG

Date _____

Owner _____ Address _____

Site Location _____ Well No. _____

Well Drilling Contractor _____

Type of Well: Percussion Drilled   _____   **Drill Hole**:   Depth ____ feet
                   Rotary Drilled   _____   Diameter ____ inches

Casing: Type____ Aboveground ____ ft   Belowground ____ ft   Diameter ____ inches

Well Protection: Grout: Type _____ Oversize Drill Hole Diameter____ inches
                            Depth ____ ft

                    Seal: Type _____ Pitless Adapter? _____

Formations Penetrated

Kind, Thickness, Water Bearing?	Level Below Grade	Diagram
_____	____ ft to ____ ft	
_____	____ ft to ____ ft	
_____	____ ft to ____ ft	
_____	____ ft to ____ ft	
_____	____ ft to ____ ft	
_____	____ ft to ____ ft	
_____	____ ft to ____ ft	
_____	____ ft to ____ ft	

Drilling Start _____   Drilling Complete _____

Pumping Tests

Static water level below grade ____ ft   Was water clean at end of test? _____
Pumping rate   ____ gpm   Was the well disinfected?   _____
Pumping level below grade ____ ft,   Is distribution system installed? _____
Duration of test   ____ hrs   Was system Disinfected?   _____

Pump Model _____ gpm _____ HP _____ Voltage_____

WELL DRILLER _____
                          Signature

William F. Albern, Sunnyslope Terrace, Ithaca, NY 14850 607-272-5077

Figure 6.5
Sample well log.

Small Water Systems

7

General

Water to be consumed by the human inhabitants of a development can have extremely adverse effects upon those inhabitants. Clean, safe water must be of paramount importance in the development of any housing project. Public water and public sewers should be incorporated into any project whenever possible. The local health department will probably insist upon it.

In the northern U.S. there is what is known as a Ten State Standard; the actual title of the document is "Recommended Standards for Water Works" and it was developed by the states of Illinois, Indiana, Iowa, Michigan, Minnesota, Missouri, New York, Ohio, Pennsylvania, and Wisconsin. The Foreword states, in part, "These standards … are intended to serve as a guide in the design and preparation of plans and specifications for public water supply systems …". There is a similar document for sewage works. The U.S. Environmental Protection Agency (EPA) has documents setting criteria for water supply and waste treatment systems.

The design of public water supply systems and public sewerage systems are extensively treated in those documents and will not be discussed in any detail here nor in Chapter 8 on Sewage Disposal. In fact, municipal water and waste treatment plants are not within the perview of this book. The design of distribution systems to connect to the municipal systems are included in those documents and in many other texts.

Small systems serving manufactured home communities and operated by the community owners, or serving small subdivisions and operated by a homeowner's association are also tightly controlled by state and local health departments. These systems are sometimes known as "private-public" systems — privately owned and operated serving the public! Finally, the criteria for the design of individual on-site waste treatment systems is also established by state and local agencies.

Any attempt within this text to describe details for the design of water supply and waste treatment systems would be redundant and presumptuous.

But, I do wish to offer a few comments in each area; here on small water systems.

Water Service

Brayton Foster discussed well systems extensively in the previous chapter. That is the most common source of water if municipal water is not available. The local health department will have great interest in any well system that serves more than a single privately owned residence. A manufactured home community or a homeowner's association will have to provide a system to pump, treat, measure, and store potable water. Figure 7.1 is a schematic of a system that should satisfy the health department with only minor changes in most instances.

Backflow Prevention

Prevention of contamination of water supplies is a constant concern of health departments and water purveyors. Water can become contaminated when a low pressure condition is created in the water mains and water from a facility "backflows" into the main water supply. Two actions that can cause a low pressure condition is a water main break and a fire department pumping water for a fire. An insecticide sprayer attached to a garden hose to control the pests around a home can be deadly. In case of a low pressure condition, the contents of the sprayer can enter the home water piping and the municipal system. Everyone always should be concerned about backflow.

There are several types of devices to prevent backflow. They range from antisiphon valves, check valves, double check valves, to reduced pressure zone backflow preventers. As with other facets of the design and construction of housing developments, this subject is well treated by state health departments and water purveyors. Contact them for the details of design and installation.

Reference

There is an excellent manual on small water systems. It is the "Manual for Small Water Systems Serving the Public correlated with National Drinking Water Regulations", Conference of State Sanitary Engineers in cooperation with Office of Drinking Water, U.S. Environmental Protection Agency.

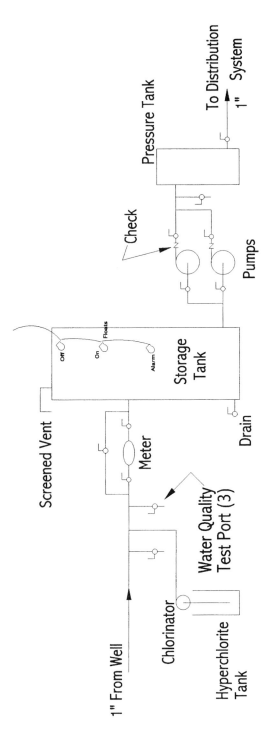

Figure 7.1
Private water storage system schematic.

Sewage Disposal 8

General

Disposal of sewage wastes from a development is most important and probably the most time-consuming design effort after the actual arrangement of the lots in the development. As noted in Chapter 7, sewage disposal is extensively addressed in government documents, thus will not be discussed in any detail here. However, as with Chapter 7, I do wish to offer a few comments regarding the disposal of sewage.

You will find that the health departments will promote the installation of municipal water systems before the installation of sewers if funding is not available for both. In Chapter 7, I stated that clean, safe water must be of paramount importance in the development of any housing project. Unsafe water can cause life-threatening epidemics. For this reason, it is not at all unusual to have municipal water available for a development but local on-site wastewater disposal.

Small Flows, September 1988 states, "The population size and distribution of homes in small communities and rural areas usually make conventional gravity sewers costly and impractical. In these settings, wastewater is usually treated by onsite systems". In the Spring 1994 issue, *Small Flows* reports that "according to the 1990 census, there were more than 25.9 million onsite systems in the United States."

The most common form of on-site system is the underground absorption trench system. Other on-site systems include sand filters, mound systems, and raised beds. The design of these systems for a home is predicated on the number of people occupying the home and the suitability of the soil to absorb the effluent from a septic tank. Not too many years ago, you dug a couple of 18-in. deep holes, dumped in some water, and measured how fast the water level dropped one inch. That's not the system today. Now we are concerned about where is bedrock, where is clay, where is groundwater, and where is it going? Again, the subject is treated substantially elsewhere.

Small Flows

Small Flows is a publication of the National Small Flows Clearinghouse at West Virginia University, P.O. Box 6064, Morgantown, West Virginia 26506-6064, 800-624-8301. It is published quarterly and reports extensively on innovative waste treatment systems and applications. The publication is sponsored by the U.S. Environmental Protection Agency.

Most, if not all, editions of *Small Flows* list a very comprehensive assortment of documents that are available from the Clearinghouse at nominal cost. The listing is preceded by the statement, "National Small Flows Clearinghouse Products List. The National Small Flows Clearinghouse is designed to help small communities reach practical, affordable solutions to their wastewater treatment problems. We offer more than 275 different manuals, booklets, pamphlets, and videotapes. These materials range from technical design manuals that detail system design, to general interest videotapes to help a small community plan for its environmental needs."

The University also publishes, but not on a regular basis, *The Small Flows Journal,* "a collection of professional papers on the study of onsite and small community wastewater issues." There is no cost to receive the publications.

Small Flows has had articles on a broad range of systems. Included have been lagoons, constructed wetland treatment systems, how to select a consulting engineer for a small community system, grinder pump systems, septic tank maintenance, sequencing batch reactors, and vacuum sewers. The Spring 1995 issue contained a listing of computer-based bulletin boards that addressed environmental issues.

Septic Tanks

As reported by the National Small Flows Clearinghouse in its September 1988 issue of *Small Flows,* Joseph R. Makuch and William E. Sharpe (in a circular published by the Cooperative Extension Service of The Pennsylvania State University) discuss Two Remedies for Failing Septic Tanks.

Small Flows states, "The authors discuss the two types of septic tank failure, efficiency and hydraulic. In efficiency failure, the wastewater is not purified sufficiently before discharge to the groundwater below the drainfield. In hydraulic failure the absorption area is unable to accommodate all of the wastewater being discharged to it. Hydraulic failure will almost always be accompanied by efficiency failure, while efficiency failure can occur by itself."

That statement may seem obvious and elementary. However, the condition is often overlooked by a designer or owner trying to reduce costs. The cleaner the effluent is from a septic tank the less clogging it is likely to cause in the disposal system secondary treatment area. Small septic tanks do not allow the effluent adequate time to settle and provide aerobic digestion.

The cost of a septic tank is a small part of the cost of developing a home in a manufactured housing development within a municipality that does not have municipal sewers. The list price of a 1000-gallon septic tank is $600, a 1500 gallon tank is $950, and a 2,000 gallon tank is $1575. Waste from a home needs time to settle. And time is needed for the bacteria to do their thing. Go up one size!

Filtration

Yes, I advocate filtration of the effluent from a septic tank. For the same reason I suggest a larger tank — distribute as clean an effluent to the absorption system as is economically possible. For the past several years all my on-site wastewater system designs have included a filter on the final septic tank discharge to the absorption system. The filter will trap most particulate and grease which may pass through the septic tank. The arrangement of a Zabel filter in a tank is shown in one of the details in Appendix 8. Figure 8.1 is a photograph of a Zabel filter.

The Summer 1995 issue of *Small Flows* reports on a study then being conducted on filtration of septic tank effluent by Tennessee Technological University. Filters manufactured by Zabel and Orenco were tested. A conclusion was that there is a significant decrease in total suspended solids (TSS) and biological oxygen demand (BOD) as a result of filtration. For further information, contact the principal investigator of the study, Larry Roberts, Ph.D., Professor of Civil and Environmental Engineering, at 605-372-3602.

Naturally, the more area provided for the effluent to be absorbed into the soil the less likely it is that there will be system failure. If clear water in reasonable quantities, like rain, is discharged to the soil, there will not be a problem. Flooding would be an example of too much hydraulic loading.

Chlorination

Although not usually required by any local code, I will always include a chlorinator with a sand filter system, because a sand filter yields an effluent. Years ago, this effluent frequently was piped to a stream or pond. Today, most effluent goes to some sort of ground absorption system. That could be an absorption trench system, a mound, or other system acceptable to the local governing authority. If the effluent from the home eventually must be absorbed by the soil, why should a sand filter be used? Although I have never had a succinct answer to the question, I advocate that the poor soil condition

Figure 8.1
A Zabel filter.

will, at times, cause effluent to reach the surface. The sand filter treatment, before the absorption system, will remove most of the objectionable organisms. However, not every time. Therefore, I will specify a tablet style, passive chlorinator be installed in the outlet pipe of the sand filter. Figure 8.2 shows a chlorinator.

Quantity of Sewage

A significant amount of my practice is in rural areas with farming operations in the vicinity. More than once a community member has addressed a public hearing with how the development may contaminate the existing wells with all the sewage that will be generated by the project.

On page 93 of the July 1990 issue of *National Geographic* there is a statement that "45,000 dairy cows … produce as much raw waste daily as a

Figure 8.2
A pasive chlorinator.

city of 980,000 people." Any development that replaces a farm housing live-stock will probably reduce the quantity of waste being generated.

Alternative Sewers

Another publication of the National Small Flows Clearinghouse, *Pipeline,* in its Fall 1996 edition, discusses alternative waste disposal systems; that is, alternatives to the conventional gravity municipal sewer and the previously discussed on-site systems. The publication states, "Alternative Sewers May Be a Good Option if …

* conventional gravity sewers and onsite wastewater treatment technol-ogies have been determined to be inappropriate or too expensive;
* the population in an unsewered area is such that there would be 50 to 100 homes or less per mile of sewer line;

★ homes are located in hilly, rocky, low-lying or very flat areas, or areas with shallow bedrock, a high water table or other site conditions that would make installing gravity sewers impractical; or

★ areas are experiencing potentially costly problems with existing conventional sewers that are leaking or deteriorating."

The article reviews four different alternative systems, all incorporating pressure sewers, defined in the article as "a small-diameter pipe into which partially-treated wastewater is pumped and then transported under pressure to a final treatment facility or to a conventional gravity sewer main." The four alternative systems are

- The septic tank effluent pump, STEP, that "consists of a septic tank to pretreat the wastewater and a submersible, low-horsepower sump pump to push wastewater through the system."
- The grinder pump system that "works something like a garbage disposal. Solid materials in the wastewater are cut up and ground into tiny pieces. All of the wastewater is then pumped out into the pressurized line."
- The small diameter gravity sewer, SDGS, that uses "gravity, rather than pumps or pressure, as the main force to collect and transport wastewater …". An SDGS system is laid at varying grades that create low spots where the effluent collects until pressure builds up to propel the effluent over the "hump".
- Finally, the article discusses vacuum sewers. "Vacuum sewers rely on the suction of a vacuum, created by a central pumping station and maintained in the small-diameter mains, to draw and transport wastewater through the system to final treatment."

The National Small Flows Clearinghouse has additional literature, available on request, on the pressure systems (see the section above on *Small Flows*).

Other Utilities

9

Routing

There are several other utilities needed to serve a manufactured home community and for a realty subdivision. And the space required for utilities must be of concern to the designer.

In a manufactured home community the utilities may be placed wherever the owner wishes. The owner can bury and dig in any location. Pragmatically, most utilities follow the roads, although I have been known to extend a sewer across lots when it made economic sense to do so. I try to avoid going directly under the space a home will occupy. Today, like in most realty subdivisions, most electric utilities are underground.

In a realty subdivision, the utility company will want easements for any service crossing private property. The homeowner grants the utility an easement, permission to maintain the underground utility. Should a problem occur, the utility can dig. Twenty years later that dig could ruin a beautiful azalea garden! As a result, every effort is made to keep the utilities in the public rights-of-way (ROW). Sometimes there is no way to avoid a sewer pipe crossing private property. When that becomes necessary, a property line is frequently followed with one half of the easement in each parcel.

Roads in a subdivision are located in the municipal ROW. Although the widths of the ROWs vary with the state and locality, typically they are 60' wide. Residential roads are often 22' wide. That leaves 38' for the utilities. Sounds like a lot, but there are drainage ditches on one or both sides that can take 6 to 10' each; there is water, fire hydrants, sewer, storm sewer, electric lines, the information cable, telephone, and natural gas lines. In some areas geothermal hot water, or central steam, may be added. The information cable wasn't around 20 years ago. What else will have to go below ground 20 years from now?

In most municipalities the subdivision and zoning regulations specify a "setback" distance for the space occupied by a home. This is the minimum distance each home must be back from the edge of the ROW. Typically, the minimum setback distance is about 30'. Also, the larger the lot size required by the zoning, the deeper will be the setback distances. Proponents of cluster housing and more open space decry some of those distances. Some have the philosophy that the home should be near the road with lots of open space in back.

However, the setbacks are part of the development of a subdivision, and I use it to control the costs of development. I will have an additional 15' of easement on each side of the ROW. Now there is a 90' wide corridor for the road and all the underground utilities. The utility companies can engineer and install the necessary facilities better and more economically, and the reduced costs can be passed on to the developer then to the home buyer. The 15' wide easement is land the home owner cannot use; it is in the setback. Yes, 20 years later, a tree may be dug up and it will be a loss. But underground failures are few and far between and the utility normally can work around most vegetation.

Fuel Oil

Just three words about fuel oil. Don't do it!

Years ago many manufactured home parks installed fuel oil systems to better serve those leasing space in the park. Also, it generated income to the park owner and eliminated unsightly storage tanks. Today, we recognize that fuel oil spills can contaminate the environment. The fouling of water wells is the most obvious. And, in any underground fuel oil system installed in a manufactured home park, there will eventually be a leak in a line. The Environmental Protection Agency has extensive regulations regarding fuel oil leaks.

Those homeowners wishing to heat with fuel oil can have an above-ground tank, properly installed and protected. It can be enclosed with material compatible to the home and park. Then, everyone is happy.

Storm Water Drainage 10

General

Storm drainage, like sewage disposal, is extensively treated by others and is only listed here to recognize that it is a significant consideration whenever land use is changed, be it for a housing development, industrial plant, recreational park, or whatever. Erosion of existing land is undesirable and runoff from one parcel of land may not cause adverse impact on others.

The United States Department of Agriculture, Natural Resources Conservation Service, formerly the Soil Conservation Service, has several guidelines that include methods of calculating runoff, systems to retain rainwater to lessen the impact on adjacent property, details to reduce erosion during and after construction, and reference to their documents is strongly suggested. Apparently, the guidelines are tailored to the laws of the states, and I am not sure if there is a version for each of the fifty states. You can obtain copies of the New York State edition of the guidelines from the Empire State Chapter, Soil and Water Conservation Society, Box 7172, Syracuse, NY 13261-7172.

Erosion Control

I normally will use this, or something similar, on my drawings for the control of stormwater runoff:

EROSION CONTROL

Conform to the New York Guidelines for Urban Erosion & Sediment Control. Obtain Department of Environmental Conservation (DEC) permit for storm water discharges prior to start of construction. During construction, install and maintain bright orange snow fencing at drip lines of all trees in construction area. Keep construction equipment away from existing trees. Construct drainage ways along property lines concurrently with road construction.

Install straw bales in roadside ditches, diversion ditches, and drainage ways as grading is progressing. Reinstall as required at the end of each work day. Install new bales, minimum 100' intervals, after seeding. Provide silt fences or straw bale dikes, in accordance with aforementioned erosion control guidelines, as required and directed, in areas being worked. Fences or dikes will be required to prevent runoff of removed topsoil which has been spread and distributed. Install rip-rap at all culverts and as directed.

Immediately after the grading of an area is completed, sow creeping red fescue at 60#/acre and annual ryegrass at 25#/acre for erosion control. Roadside ditches shall be particularly promptly treated. Install straw/fabric silt fences in drainage ways to prevent erosion.

File Notice of Intent for Storm Water Discharges Associated with Construction Activity.

Operations 11

General

A manufactured home community is managed by the community owner who is charged with operating the community. However, realty subdivisions that have homeowner's associations also have an operating function, especially if the homeowner's association owns and operates a utility, community center, swimming pool, etc.

Again, this is a subject extensively treated elsewhere. While this book is primarily concerned with the *planning, design, and construction* of developments, a few of the appendices contain information that can be helpful with manufactured community operations. Appendix 6 is a typical set of rules and regulations for the operation of a community, including the lease. Appendix 7 is an eviction notice.

Additional and quite complete operations information for manufactured home communities is available from these and many other sources:

- The manufactured housing association in each state
- The Manufactured Housing Institute, 1745 Jefferson Davis Highway, Arlington, VA 22202, 703-979-6620, 800-505-5500, http://www.mfghome.org
- The Institute of Real Estate Management, National Association of Realtors, 430 North Michigan Avenue, Chicago, Illinois 60611-4090, 312-329-6000
- American Planning Association, 1776 Massachusetts Avenue, NE, Washington, D.C. 20036, 202-872-0611, http://www.planning.org
- Manufactured Housing Global Network, http://mobilehome.com

Accredited Community Manager

This is the one area of operations that I shall address briefly because I believe it can improve the environment of manufactured home communities greatly.

The Manufactured Housing Educational Institute, an affiliate of the National Housing Federation, has developed a national comprehensive education and accreditation program designed to enhance the professional standing of community managers. A program of this type can improve the livability of the nation's manufactured home communities.

The Manufactured Housing Educational Institute can be contacted at Suite 106, 3015 Williams Drive, Fairfax, VA 22031, 703-558-0400.

Amenities 12

General

Chapter 4 detailed the layout of factory-built housing developments. It asserted that curving roads create a more pleasing place to live. The esthetics of the development are much improved when every lot is not the same shape and size. Within reason, the more varied the plots, the more attractive is the housing.

A variety of lot shapes is only one of the many amenities a developer can offer. Others that can attract the public to a particular development include community buildings, recreational parks, trails, open natural areas, local convenience shopping, a nearby elementary school, and athletic facilities. A creative developer should realize others.

A community facility can be a major attraction for people to live in a specific development, be that development a realty subdivision or a manufactured home community. The facility can be as simple as a one room multipurpose building for use by the residents for meetings, card games, girl scout and boy scout meetings, day care, etc. Or, it can be a much more elaborate facility with a convenience store, swimming pool, bowling alley, or whatever the development can support.

A very small amenity with great returns is the central postal boxes as shown in the photograph of Figure 12.1. By centrally locating the post boxes in a drive off area at the entry to the development, the unsightly mail boxes are removed from each lot, the mail truck is not adding to the traffic and people are not stopping at their front yard mail boxes and stopping traffic. They are a win-win amenity.

Any facility must be able to be supported by the residents, and the developer must consider the economics carefully, lest the facility become more of a liability than an asset. In a subdivision the cost of the facility must be apportioned among the lots being offered for sale. A homeowner's association can be created to deal with the continuing utility and maintenance costs. In a manufactured home community, the lot rentals will need to cover the capital costs of the facility and the continuing operating costs.

Figure 12.1
Centrally located mailboxes.

Zoning ordinances are often a hindrance to providing a community facility. An area of the municipality may be zoned residential permitting a subdivision. However, the definition of that residential zone may prohibit a non-housing unit. It may seem so simple, and it should be, to have the definition changed to permit the community facility. But municipal boards are typically reluctant to make such a change. At the very least, it is going to cost the developer time and money to appear before one or several more meetings of the municipal board to persuade the board to allow the mixed use in the zone. Should the request be for a convenience store in addition to a simple community building, the municipal board's resistance will probably be exponentially increased.

Sports facilities are another item a developer can use to attract people to a particular community. Anything is possible, from a simple basketball back-board to a swimming pool complex. For information on the physical sizes required for various sports, the American Institute of Architects, *Architectural Graphics Standards* by Ramsey and Sleeper (John Wiley & Sons, New York) will be found to have most of the necessary information.

Any amenities a developer wishes to include must be known to the person designing and laying out the development early in the project. Many amen-ities, as just discussed, will require space within the development. It is harder to position the amenity ideally after much layout has been done. Better to begin with the requirement as part of the project.

The Drawings 13

General

The vehicle that moves a development from the owner's dreams and then through planning to construction is the set of drawings evolved during the process. The owner may very well have started the process with a sketch of his desires, or those desires may have just been transmitted orally to a designer. The size of the project and the municipality in which it will be constructed can influence the extent of the drawings. I am sure that in rare instances, a sketch by an owner on "the back of an envelope" has been used to construct a development. In most cases, though, for a development to proceed, professionally prepared drawings will be required by the approving jurisdictions. (See Chapter 3.)

The Professionals and the Drawings

The Surveyor

For a realty subdivision a licensed land surveyor will be needed to describe properly the property to be subdivided. The metes and bounds of each piece of property are required to be indicated on a drawing/survey. Also, a topographic survey and boundary survey of the entire property normally will be required. A Surveyor's Plat is the first drawing in Appendix 10.

For a manufactured home community, a surveyor may or may not be party to the project. For a site with about 50 homes, a tax map and the United States Geological Survey (USGS) topographic maps will probably satisfy the project needs. Large developments may need more precise information. It is impossible to state when a surveyor will or will not be required. The community size, municipal jurisdiction, land features, etc. will dictate the need for a surveyor.

The Engineer

A licensed engineer will be required to prepare water supply and sewage disposal details for the development, in addition to the design of the roads and the drainage patterns.

Over the years I have developed a few "standard details" that I repeatedly use in the engineering construction drawings. Not surprisingly, they are mostly on sewage disposal. An assortment of them is included in Appendix 8.

Additionally, the developments in which I participate are normally not very large. As a result, I have not found a "specification manual" to be a mandatory item. I have been able to include the necessary specification information as part of the drawing plans, as part of the details, or as a short listing of "notes" on the drawings. Appendix 9 is a sampling of my frequently used notes for the plans. Please understand that the notes listed have conflicting sentences. That is because all the notes are never used; appropriate notes are selected and arranged into a readable paragraph format for each individual project. Additional notes are normally needed specific to the immediate project. The notes are then added to the drawings, see drawing 4 of 6 in Appendix 10. The file of "standard details" and the file of "notes" act as mini-checklists for me when I am drawing a project. By checking those two files I am reminded of details that need to be added to the current drawing.

Appendix 10 contains the six engineering drawings for a realty subdivision project plus the Plat as previously mentioned. Some of the details, especially the small text, are difficult to read due to the necessary size reduction to fit the dimensions of these pages, but the titles of the details are legible. This is a good example of the extent of a set of engineering drawings for a project. Appendix 11 contains the engineering drawings for a manufactured home community.

The Architect

An architect is needed for a project if there will be structures as part of the development, other than those associated with the water, sewer, road, and drainage systems. The architect would develop the plans for the buildings to be constructed. Please remember that for a Housing and Urban Development approved (HUD) manufactured home, architectural plans are not required.

The Lawyer

A lawyer would prepare the deeds for a realty subdivision and provide other legal advice, primarily to the owner.

Others

Additionally, other specialists may be necessary for a specific project. Someone with groundwater expertise may be required, an archeologist if American Indian artifacts are expected or found, a planner to work with the municipal planning board, etc.

Please note, I charged none of the preceding with the task of laying out the configuration of the development. Anyone can do that. However, a person without experience in lot layout will probably produce a drawing that will not readily be accepted by the local municipality. I have done most of the lot arrangement designs for my projects. But there is always assistance from others in the field to improve an arrangement. The owner and municipal planning staff will frequently suggest changes to produce a better product.

Factory Construction 14

Construction in the Factory

You will probably be surprised with the methods for constructing factory-built housing in the manufacturer's plant, as I was. Manufactured housing is built from the center out! The tub is placed on the floor before the walls are erected. When you think about it, it becomes obvious. Why have a wall in the way of putting a tub in place?

Figure 14.1 is the beginning of construction of a factory-built home. The floor is being framed at this first step. HUD Code approved manufactured homes and modular homes travel the same assembly line in this plant.

The second station on the assembly line is the installation of utilities under the floor (Figure 14.2). The floor from Figure 14.1 has been lifted by a crane to allow workers to install the required plumbing-heating-electric systems underneath.

In Figure 14.3, note the hose on the left. It connects to a compressed air system. Electric outlets are visible on the right. There is also lighting available for the workers. Those kinds of service, plus no rain days, are what decreases the cost of the factory-constructed home.

The floor has now been moved to the next station, Figure 14.5. The tub is the first item to be mounted. Figure 14.6 shows sheetrock being nailed to a wall frame. The nailer drives the nail and dimples the sheetrock in one action. The wall is now ready for spackling.

Figure 14.8 may be a bit difficult to understand. A home has been moved to the next position; a catwalk hides the upper portion of the home. The catwalk can be raised and lowered. It allows workers access to the upper portions of the homes without having to contend with ladders.

The ceiling/roof is being constructed in a parallel portion of the plant. Sheetrock for the ceiling of the home is laid on a flat table. In Figure 14.9, the roof trusses are constructed above the sheetrock. Notice how the roof pitches for a bit and then is flat. More in a minute.

The joint between the roof trusses and the sheetrock is "foamed" (glued) together (Figures 14.9 and 14.10). No nails are used to connect the sheetrock

Figure 14.1
A manufactured home floor being assembled.

Figure 14.2
The floor from Figure 14.1 has been lifted to the next station.

to the roof trusses. Any voids due to warped or twisted trusses are filled with the foam. The ceiling sheetrock is flat.

Figure 14.11 is a close-up of the trusses. Note the round connections in the joining plates. It joins the pitched portion of the joist with the horizontal

Figure 14.3
Installing flexible ductwork.

Figure 14.4
Installing the plumbing.

as mentioned in Figure 14.9. That is a rotatable joint. The horizontal portion is raised after delivery to the site, to follow the pitch of the other portion of the truss. This allows the home to be transported on highways and, more importantly, under highway bridges.

Figure 14.5
Building on the floor.

Figure 14.6
Power nailing sheetrock to a wall frame.

Figure 14.7
Moving a prefabricated wall partition to the floor of the home.

Figure 14.8
The catwalk for access to the roof.

Figure 14.9
Roof trusses.

Figure 14.10
Foam joining trusses and sheetrock.

Figure 14.11
Rotatable truss joint.

Figure 14.12
Spackling a home.

Figure 14.13
Installing the lighting in the ceiling.

Figure 14.14
Electric service panel.

Figure 14.15
Plumbing service for a washer.

Figure 14.16
Kitchen cabinets being installed.

APPENDIX 1

Contributors

Brayton Foster

Brayton Foster has a degree in Geology from St. Lawrence University and a Masters in Geology from the State University of New York (SUNY) at Buffalo. He is an independent consulting geologist with experience in oil and gas exploration, drilling supervision, and well testing. He acted as on-site representative for the New York State Energy Research and Development Authority (NYSERDA) during drilling and pumping for a geothermal site in Auburn, NY. He has been geologist for Central New York water well problems, pump testing, water resource evaluations for subdivisions and manufactured home parks. He has experience with fracturing of water wells. Mr. Foster has publications on Models of Dikes, Analysis of Oil and Gas Resources in Allegany and Cattaraugus Counties in New York, the Ithaca Kimberlites, and the Kimberlites of the Finger Lakes region. He is a member of the Central New York Association of Professional Geologists and other organizations.

Jim Ray, Sr.

Mr. Ray was active in the New York State Manufactured Housing Association, and his legislative committee was successful in getting legislation passed that put in place one of the first new home warranty laws in the nation. A tenants rights bill was also implemented that maintained the most important private property rights of the community owner. Mr. Ray was appointed by the New York State Governor to represent the mobile home industry on the New York State Building Code Council at a time when the state was implementing a new construction and installation code for mobile homes.

APPENDIX 2

Appendix 2 is reprinted from *Managing Mobile Home Parks* by Stephen G. Papas, CPM®. (Readers should seek competent professional advice before using any of the information provided herein.) Copyright 1991 by the Institute of Real Estate Management, 430 North Michigan Avenue, Chicago, Illinois 60611. Reproduced by permission of the publisher. For a catalog of publications and education offerings or to obtain a copy of *Managing Mobile Home Parks*, write to the address shown or call 312-661-1953.

Laws and Zoning

"But in this world nothing can be said to be certain, except death and taxes."
—**Benjamin Franklin**

Franklin might also have added laws, rules, regulations, codes, permits, licenses, ordinances, and statutes (to say nothing of bylaws, acts, mandates, directives, and edicts). Regardless of what they are called, it is important to remember that interpretation of a particular law is based on the law's intent rather than the actual words used.

There are many legal requirements that affect the development, expansion, and operation of a mobile home park, and managers and developers must be aware of all of them because there are legal implications to every action taken while conducting a business. Naturally, it is not possible (or appropriate) to explain the fine points of all of the laws at the state and local level in a book that focuses on management. There are, however, similarities among the laws that affect mobile home parks, and this chapter offers the park manager a starting point for understanding them.

Laws in General

The appeal of real estate investment can be explained in part with a few words — property rights and human rights. Both are guaranteed by the Bill of Rights and the U.S. Constitution. All laws that govern real estate in the United States are based on these two documents, which have influenced state constitutions, laws passed by the U.S. Congress and by state legislatures, rules and regulations promulgated at the federal and state level, local (municipal and county) ordinances, court interpretations (case law), and common law. In general, the laws that affect mobile home park operation can be categorized as follows:

- National or federal laws
- State or municipal laws
- Zoning ordinances and development and building codes
- Operational requirements

Laws promulgated by the U.S. Congress are applicable nationwide. Examples include fair housing laws, energy conservation laws, gas pipe corrosion laws, environmental laws (air and water pollution, solid waste disposal, noise, use of solar energy, etc.), the Occupational Safety and Health Act of 1970, and urban renewal and housing programs. The impact of federal law on mobile home parks and park operation is varied. Laws regarding fair housing, protection of the environment, and energy conservation can also be developed at state and local levels and may have a direct impact on mobile home parks. State laws that affect mobile home parks include landlord-tenant laws, rent controls, contract laws, and laws regarding taxation. Some local jurisdictions may develop more stringent versions of these types of laws. Zoning ordinances and development and building codes are primarily local in origin. These laws may do such things as restrict land use, establish health and safety standards, or initiate a no-growth (or limited-growth) policy. Such legislation may limit development on a floodplain or require that people who build on floodplains carry flood insurance. While some manufactured housing may be subject to local zoning or deed requirements, the construction of all mobile homes in the United States must comply with the same building code — the National Manufactured Home Construction and Safety Standards. Operational laws include permits (swimming pools, special events, etc.), business licenses, guest registration, gaming licenses (e.g., bingo), and some employment laws. These laws are also local in origin.

Laws that regulate the workplace, however, are in effect at all levels. Federal laws include the Fair Labor Standards Act which sets minimum wage rates and the Equal Opportunity in Employment Act which prohibits discrimination. States regulate work hours and payment for overtime as well as unemployment insurance. Local jurisdictions may issue various ordinances or impose taxes that apply to the employer. The range and varieties of all these different laws is beyond the scope of this book. The manager should be familiar with all the laws at all levels that may in any way affect the operation of the mobile home park he or she manages.

Brief descriptions of the laws or legal requirements that are most likely to affect mobile home park management are presented in the following sections. Some laws will be described in greater detail in later chapters. Landlord-tenant laws and zoning are major issues, and these are detailed in separate sections later in the chapter.

Specific Legal Requirements

Fair Housing Laws

Local, state, and federal fair housing laws exist to prevent discrimination in housing. One such law is the Fair Housing Amendments Act of 1988. The Amendments Act was designed to protect families with children and people with handicaps. In general, federal fair housing laws prohibit discrimination based on race, gender, nationality, religion, age, family status, and disability. Local laws may exist to protect additional classifications of people (e.g., housing discrimination based on lifestyle is prohibited by some local laws). Different forms of subsidization exist to promote fair housing. Like federal fair housing laws, nationwide subsidy programs are administered by the U.S. Department of Housing and Urban Development (HUD). Fair housing laws have far-reaching consequences; even construction may be affected (e.g., construction plans may have to be altered for handicapped residents).

Environmental Regulations

These laws — federal, state, and local — outline such things as the acceptable limits of air, water, and noise pollution, water quality, and land drainage (sanitary and storm sewers). Prior to mobile home park development, an environmental impact statement (EIS) may be required under federal or state law (e.g., under the National Environmental Policy Act of 1969, an EIS is required for federally supported developments). An EIS is an analysis of the anticipated effects of a development or action on its surroundings. Such an analysis might reveal impediments to development (e.g., the presence of underground tanks or toxic waste).

Energy Conservation Laws

These laws are usually in place at the local level and enforced through the local utilities. Most required energy conservation procedures are designed to save money over the long-term, although they often require a greater initial expense.

Landlord-Tenant Law

This is the basic law that governs rental residences, including mobile home parks. Landlord-tenant law is enacted on the state level and there are variations in every state. There may also be local ordinances that are applied as extensions of landlord-tenant law. In general, these statutes and ordinances deal with the occupancy of a space and cover such subjects as evictions, forcible detainers, abandonment, and leases (discussed in detail later in this chapter). There has been a movement by the states in recent years to create a landlord-tenant code that applies specifically to mobile home parks, separate from regular residential statutes.

Tax Laws and Regulations

State and local taxes on mobile homes take many forms. The manager of a mobile home park should understand his or her state and local tax picture and know the procedures for filing tax protests.

In some states, mobile homes are considered personal property (personalty) rather than real property (realty); in such states, property taxes are paid only on mobile home sites, not on mobile homes. In other states, mobile homes are treated as real property, and local property taxes are levied accordingly. Property tax is an ad valorem tax because it is based on the value of the thing being taxed. Property tax is determined by applying the local tax rate to an assessment of the property value (the assessment for tax purposes is done by the local government).

Some states "tax" through licensing fees, and there are those states that regulate mobile homes through vehicle registration laws. Rental tax may also be applied to mobile homes at the state level; the collection of such a tax is usually a responsibility of the park manager.

Protesting Taxes   A great deal of money can be saved by the manager who knows the local procedure for protesting a tax assessment on his or her park. In order to protest, it usually is necessary to file a petition with the county assessor by a certain date. By law, the assessor must respond to such petitions within a set number of days. If the adjustment obtained from the assessor is not satisfactory, the protester can usually file for a hearing before the appropriate body (e.g., a Board of Equalization). Generally, the majority of such appeals is resolved through this means. If the results are still not satisfactory, however, the manager can file for a final hearing with the state or local board that adjudicates property tax appeals. The definitive source of appeal, of course, is the courts.

Rent Control Laws

Usually local in nature, rent control laws attempt to provide for "affordable housing" by regulating the amount that can be charged for rented space. While these laws may limit the amount of rent that residents have to pay, rent control may also have some negative residual effects (e.g., the quality of affordable housing may decline because owners have fewer incentives to make repairs). It is mandatory for any mobile home park developer, owner, or manager to be aware of local rent control laws — and these law are changing all the time.

Land Use Laws

Primarily local, these laws dictate allowable land uses through such restrictions as zoning and related ordinances (discussed in more detail later in this chapter). These ordinances limit types and sizes of structures, population densities, and building placement (e.g., setbacks). Land use is also regulated via deed restrictions or covenants.

Local Building Codes

These codes establish minimum construction and safety standards for buildings (including mobile homes). Such codes often require the purchase of permits.

Health and Safety Requirements

While there are federal and state laws regarding health and safety, the laws or codes that have immediate impact are those local requirements regarding fire prevention, swimming pool and restroom cleanliness, refuse disposal, internal and external security, sewer and septic tank uses, etc. The following (or similar requirements) will usually apply.

Health Requirements   Inspection of swimming pools, washrooms, and kitchens for compliance with local health regulations is usually a function of county or city health departments. Certificates of inspection may be issued, and these may have to be displayed prominently.

Refuse Disposal   Requirements may call for storage in watertight, rodent-proof containers, sufficient in number and capacity to prevent overflowing. Laws that restrict the storage, collection, and disposal of refuse prohibit the creation of such things as health hazards, insect or rodent breeding areas, and air pollution. As more and more landfill disposal sites are closed and the popularity of recycling increases, legislation may limit what can be disposed as landfill. These laws are usually initiated at the state level and may be further defined by local ordinances.

Fire Safety   Adequate fire protection is a must. To meet these requirements, the manager must keep several points in mind. First of all, a mobile home park is subject to the rules and regulations of the local fire authority. The park should have standard fire hydrants located as required by local ordinances; locally approved portable fire extinguishers should be kept in all buildings and at other appropriate locations, and these should be properly maintained. In addition, all parks should be kept free of litter, accumulated rubbish, and other flammable materials.

Inspections   A mobile home park manager should be aware of required inspections. Local enforcement agencies may inspect a mobile home park to be sure it complies with state laws and regulations. They may inspect a mobile home site when a new home is moved in, and check utility connections, stairways, storage of combustible materials, new plans, alterations or repairs, etc. During normal tours, they will cite violations that the owner or manager may not be aware of as well as those that involve resident obligations. In the case of a park that has a history of violations, inspections will be made without an appointment.

Installation inspections and other such examinations protect the resident against accidents and hazardous conditions and protect the owner against lawsuits and inferior work by subcontractors. Inspectors can provide a valuable service to park owners and managers, so they should be treated as allies, not enemies. A positive attitude is important when dealing with them.

Employment Laws

These include both federal and state laws. The Fair Labor Standards Act (often called the federal wage and hour law), state Workers' Compensation and Employers' Liability Acts, and the Occupational Safety and Health Act of 1970 are prime examples. Respectively, these statutes establish a minimum employee wage, provide for the financial security of employees who become ill or are injured on the job, and ensure safe working conditions. The Federal Insurance Contributions Act (social security) and the Federal Unemployment Tax Act provide income, respectively, when employees retire and when their jobs are terminated.

Permits

Local permits may be required for everything from towing, awnings, and skirting to tie-downs, landscaping, and utility connections.

Business Licenses

In some states, mobile home park operators must have an operating license similar to that required for a hotel or motel. Separate licenses may be necessary if there is a store, gas station, or mobile home sales agency or dealership within the park. (Mobile home dealerships are regulated by a separate set of laws beyond the scope of this book.)

Registration of Guests

The registering of guests, whether they are residents or nonresidents, is required in some states and is considered extremely important in the case of travel trailers. The major concern is the impact of increased population density on water and sewerage systems. Basic information usually includes at least the following.

- Name of the resident or visitor, the assigned park lot number, and the party's former address
- Make, model, and license number of each mobile home, vehicle (automobile), or travel trailer
- Dates of arrival and departure
- Who to contact in case of an emergency

Miscellaneous Legalities

Examples of other legal requirements are the posting of warning and advisory signs, limitations on bingo or other games of chance (gambling laws), frequency of corrosion inspections (i.e., of gas lines), requirements and limitations on contracts and licenses, etc.

Other Aspects of the Law

There is a spirit as well as a letter of the law. Because of this, a law or regulation usually can be altered if the change will make mobile home park standards more responsive to the needs of both local communities and park owners and residents. While standards that do not comply with the law's minimum requirements generally cannot be adopted, higher standards may be set. For example, if state law requires a certain minimum distance between mobile homes, a park owner may establish a greater (but not a lesser) distance as the park standard. Any differences from the law must be stated in the written Rules and Regulations of the park, otherwise state standards prevail. Regardless of the standards that are applied, enforcement of park rules ultimately depends on the owner or manager.

Landlord-Tenant Law

As stated earlier, every state has some variation on landlord-tenant law. It is a body of laws that governs residential rental transactions and provides guidance for operating within local laws. Often there are specific laws for mobile home parks. Although many mobile home park owners do indeed rent mobile homes in the same way apartments and houses are rented, the major type of tenancy is one in which the mobile home owner leases the lot or pad on which his or her mobile home is installed. It is this type of rental relationship between resident and park owner that may require unique applications or interpretations of landlord-tenant law.

The purpose of a landlord-tenant law is to protect the health, safety, and investment of both park residents and park owners. The law sets forth standards for the physical features and operations of parks and regulations for carrying out the law's intent. Landlord-tenant law is highly complex and can be easily misunderstood or misinterpreted. The park operator or manager should study his or her local version carefully; understanding it will reduce legal and administrative problems.

Some Applications of Landlord-Tenant Law to Mobile Home Parks

The following discussion of landlord-tenant laws is intended only to identify some of the basic park management procedures that are often prescribed by the state — obviously, it is crucial for the owner and manager of any mobile home park to be aware of the landlord-tenant law in their state. In a book such as this one, it is not possible to identify all the aspects of the law that apply universally because of the variations that evolve at different jurisdictional levels; therefore, competent legal advice should always be sought when specific problems are to be addressed.

Landlord-tenant laws applied to mobile home parks typically cover five discrete areas.

1. Legal principles and landlord-tenant obligations
2. Rents, the rental agreement, statements of policy, rules and regulations
3. Noncompliance and termination of tenancy
4. Termination of the park
5. Hearing officer

The first three parts will be discussed in some detail in the following sections, citing aspects that may be unique to mobile home parks. The remainder are treated only briefly because they usually require specific advice of legal counsel. Many of the issues described here would be dealt with specifically and appropriately in a lease document (rental agreement) or in the park's published Statements of Policy and/or Rules and Regulations.

Basic Legal Principles

The law usually applies only when mobile homes are placed on rented spaces. It usually does not apply to RVs, "travel trailers," or to the combined rental of a mobile home lot and a mobile home. The law may not apply to a mobile home and mobile home lot if both are owned by the same person.

Landlord Obligations

The landlord may be required to provide copies of the state landlord-tenant law to all residents; to comply with all relevant city, county, and state codes (including those affecting health and safety); to furnish outlets for electrical, water, and sewer services; and to provide prospective residents with all the information necessary to make proper use of the services. There may be requirements for notification of residents when a manager's services have been terminated, and there may be limitations on the landlord's right of access to a mobile home owned by a resident. A prospective resident probably cannot be denied a rental space unless his or her mobile home is incompatible with those already in the park or the prospect does not meet other requirements for tenancy (e.g., sufficient income). Requirements for tenancy may be stated in he park's Statements of Policy and its Rules and Regulations.

Resident Obligations

As a protection for the landlord, managers may require residents to give notice a specific period of time before vacating the rented site. The law may set limits for these notices. Residents may be denied the right to sublet without the manager's written approval. Subletting without approval may lead to penalties or eviction. Specific resident requirements are usually included in the lease or rental agreement (ultimately, resident behavior must be pursuant to lease terms).

Rent

A landlord has the right to collect rent without demanding or serving notice, and a minimum notice period for announcing future rent increases may be required (usually a written notice is necessary). If the landlord is to collect for utility charges, there may be a requirement for all units to have separate meters (e.g., 100 spaces must have 100 meters). Also, the landlord may be restricted from turning off any resident's utilities — even if a resident is delinquent in payment. The landlord may be permitted to charge a guest fee, in which case a definition of "guest" must be established. The law may allow the landlord to charge a penalty fee for late payment of rent, although some limitations usually apply (e.g., minimum grace periods and maximum late fees may be specified).

Rental Agreement (Lease)

A rental agreement must be in writing and cover a specific period of time, although the law may not set any restrictions on the period of time covered by a rental agreement (provided both parties are in accord). An agreement may not be required for established residents where no such agreement was previously required, unless requested by either the resident or the landlord. (This may be regarded as month-to-month tenancy.) In addition, the landlord may be required to provide prospective residents with copies of the Rental Agreement, Statements of Policy, Rules and Regulations, and the mobile home residential Landlord-Tenant Law. When a written rental agreement expires, the law may provide for tenancy to be on a month-to-month basis.

Statements of Policy

A copy of the park's written Statements of Policy should be provided to the resident when a rental agreement is executed. They usually cover broad aspects of park life, including (1) classification of the park and occupancy limitations; (2) the method of determining changes in rental rates; (3) size and specifications of mobile homes permitted in the park; (4) improvements that must be made to resident's mobile homes as a condition of tenancy; (5) the expected life of the park before a change in use is made; and (6) conditions under which the residents may exercise the right of first refusal to purchase when the park is sold, if such a right is given. In addition, there may be a specified period of notification for the announcement of any changes or additions to the Statements of Policy.

Rules and Regulations

The law may require adoption of written Rules and Regulations covering residents' use and occupancy rights. Rules and Regulations usually are enforceable only if (1) they promote the convenience, safety, or welfare of the residents or premises, preserve the landlord's property from abusive use, preserve or upgrade the quality of the park, and fairly distribute services and facilities of the park among all residents; (2) they are reasonably related to the purpose for which they were adopted; (3) they apply to all residents in a fair manner; (4) they are sufficiently clear so that they fairly inform the resident of what must or must not be done to comply with them; (5) they do not permit the landlord to escape his or her obligations; and (6) the prospective resident has a copy of the Rules and Regulations before he or she enters into a rental agreement. Even if it is not required by law, the Rules and Regulations should be a part of every resident's lease.

Residents may be protected from having to comply with certain substantial modifications of the Rules and Regulations that originally applied to them (i.e., at the beginning of their tenancy). For example, if a park initiates a policy requiring everyone to install awnings, only new residents may have to abide by the policy.

Park rules must be written with great care and forethought because they can have many legally enforceable consequences. A few major points to consider are:

- Rules and regulations should deal with a resident's conduct in regard to the appearance of the park and the health and safety of all its occupants. Rules regarding social conduct, beyond these areas, should be based on common sense.
- Rules must be reasonable and enforceable.
- Rules must be in writing, referred to in all rental agreements, and signed by residents and landlord/managers.
- Rules are not requests. These statements should be as brief and concise as possible and provide a reason for each rule.
- Rules should not cover general information or basic rental provisions.
- Residents must be given written notice of changes in the rules, but changes should not be sent out piecemeal. If one rule is altered, residents should be notified of the change and sent a complete new set of rules.
- Management should meet with residents who request an opportunity to discuss changes in rules.
- A written record should be maintained of delivery of all notices.

Noncompliance by the Landlord

When the landlord does not follow the terms of the rental agreement, the resident may be permitted to notify the landlord in writing, specifying the landlord's acts and omissions. The law may provide for the landlord to be given a time limit for remedying the situation. If the landlord does not act within the specified time frame, the resident may be allowed to terminate the agreement. In addition, the law may give the resident the right to correct a defect and deduct the cost of repair from the rent. (There may be other remedies available to the resident, and there may be requirements the resident must meet in relation to landlord noncompliance.) Some residents have taken owners and/or managers to court for failure to maintain the park. It is important to be aware of the decisions made in such litigation; case law interprets existing laws and sets the standards for compliance. Nevertheless, if the landlord or manager maintains the property, landlord noncompliance is less likely to be an issue.

Noncompliance by the Resident

When a resident does not pay rent, the landlord may have the right to deliver a notice specifying that rent is delinquent (past due). If payment is not made within a specified period of time, the landlord may be able to terminate the rental agreement. In the case of a resident not complying with health and safety rules, the landlord may be permitted to issue a notice that sets the amount of time allowed for remedying the situation, and if the situation is not remedied, the rental agreement may be terminated. When a resident inflicts bodily harm on anyone in the park, the landlord may be able to terminate the rental agreement immediately (upon the resident's receipt of a written notice).

The landlord may be required to specify the reason(s) for terminating a tenancy for noncompliance. The resident must be given a written notice of the landlord's intention (according to and in the form required by the law). Specific facts relating to the termination should be included (e.g., date, place, and circumstances). In particular, repetition of documented offenses may lead to termination of the rental agreement, usually in a specified amount of time after a written notice has been given. The landlord may be restricted from withholding services provided to a resident, regardless of the reason (e.g., turning off utilities for nonpayment of rent).

In any state, the definition of abandonment and the appropriate actions for the landlord to take when abandonment occurs are spelled out in landlord-tenant law. These laws vary widely. The landlord may be able to accept full or partial payment of rent and still proceed with the termination of a rental agreement.

Termination of Tenancy

A resident may be able to terminate the rental agreement without prior notification if he or she is a member of the United States' armed forces and receives reassignment orders. If, after the expiration or termination of the rental agreement, a resident still remains in possession of his or her rental space without further consent of the landlord, the landlord may terminate the tenancy by taking legal action. The landlord may not be allowed to interrupt any utility services if the resident still occupies his or her mobile home. Usually, utilities can only be stopped if the resident abandons or otherwise surrenders the rented space. If the rental agreement is terminated, the landlord may claim possession of the rental space, but a separate written claim may be required if payment is sought for any actual damage that may have occurred. When a resident refuses to allow lawful access, the landlord may be allowed to terminate the rental agreement and recover actual damages.

The death of a resident may not preclude the survivor of a joint tenancy or covenant from enjoying all prior rights and privileges and being responsible for all prior liabilities. Thus, the resident's estate may be subject to the terms of the rental agreement. In some states, the survivor of a deceased resident has the right to cancel a rental agreement by giving written notice to the landlord.

Termination of the Park (Change in Land Use)

A period of resident notification of a change in land use may be specified in the law, and such notices are usually required to be in writing. An attorney should be consulted about the landlord's liabilities in the matter of resident relocation expenses. It is usually not considered a change in land use if a resident is moved to another space in the park at the landlord's expense.

Hearing Officer

Some jurisdictions provide a hearing system for resolving landlord-tenant disputes. The qualifications and appointment of a hearing officer are usually stated in the law. A petition for a hearing that involves violations of a mobile home park landlord-tenant law may have to be in writing on a locally approved form. After the hearing is concluded, the hearing officer may be required to issue a written order within a specified period of time. The resulting order is binding on all parties unless a rehearing is granted.

Action in the Courts

Success in court depends on (1) explicit understanding of procedures as outlined in local statutes and (2) creation of what is sometimes called a "paper trail" attesting to the fact that the correct procedures have been followed. For example, if proof of having delivered a notice is required, a witness should accompany the park representative and sign the notice as a witness (or the witness can be brought to court). If the U.S. Postal Service is used, notice should be sent by certified mail, with a return receipt requested. If the notice is delivered by a process server, an affidavit swearing that the notice was served will be returned to the sender. Such notices are required for a variety of infractions of lease terms, but eviction is perhaps the one most likely to lead to court action.

The initial petition to evict should include a provision for a Writ of Restitution, which will allow the manager (or other park representative) to have the sheriff physically remove a resident in the event that he or she does not leave the park grounds within a specified time following a ruling in favor of the park.

If the resident-defendant appears in court, any questioning should be done under oath and no opportunity to get information should be missed. Questions should concern address, employment, bank accounts, social security number, nonexempt assets, and other items that will establish the defendant's ability to pay what is owed for back rent, legal fees, damages, and interest. Unless there are specific provisions in the rental agreement that state otherwise, the landlord probably will not be able to collect legal fees on an eviction proceeding heard by default (that is, if the resident-defendant does not appear).

If the defendant objects to the questions, the landlord can stress to the judge the important of these things (and his or her right as landlord to know them) in order to establish a case for collection.

The judge will usually write and sign the judgment while the parties are in court. After the hearing, a copy of the judgment can (and should) be obtained from the court clerk. (It may be useful if the landlord is forced to summon the police prior to the resident-defendant's move.)

Representation of the Park

Representation of the mobile home park (landlord) in court must be by a qualified individual. If the park is not a corporation, an owner or part-owner can do this. A corporation is a separate legal entity, and unless there are specific statutory provisions, it may be necessary for a corporation to be represented by an attorney.

Self-Representation

If the owner or manager (instead of an attorney) represents the park in court, three copies should be made of everything to be introduced into evidence — one for the judge, one for the defendant, and one for the landlord. If the document is a certified letter, the green U.S. Postal Service card should be stapled to it.

A WORD TO THE WISE: Never interrupt the judge — especially if what is being said is in one's favor.

Zoning and Rezoning

Whether expanding an existing park or developing a new one, the park owner or developer will have to acquire the necessary approvals beforehand. These are obtained through the land zoning and/or rezoning process of local government. It is important to keep in mind that this legal process is political in nature. Because it is political, it has few absolute rules and many players, a combination that can lead to considerable expense.

Ultimately, it is most important for the developer — rather than the manager — of a mobile home park to understand the nuances of zoning. For those managers who have no "hands-on" experience with development, an awareness of zoning issues can be helpful.

What is Zoning?

Zoning has often been referred to as the best known, most used, and least like growth management tool. The first zoning resolution in the United States was adopted by New York City in 1916, and since then, it has generally set the standards that have been followed by all state and local jurisdictions.

The original purpose of zoning was (and still is) to reserve land for specific purposes. Zoning ordinances can promote orderly development and prevent conflicting uses of the land in a given area. In general, zoning reserves certain parcels of land for less-intensive land uses (residential) while other land is allowed for more-intensive uses (commercial and/or industrial). Often associated with city planning, zoning — as the name implies — divides a community into zones or districts, with regulations governing such things as land use, building height, and project density.

This is not to say that such regulations are cast in stone. Again, it is important to understand zoning in light of the political process. In parts of the United States that are undergoing development, land use patterns are still being formed, and their governing ordinances are intentionally vague. On the other side of the coin, zoning is often used as a protectionist mechanism to maintain the status quo in areas that are already highly developed.

Zoning need not be perceived as a "necessary evil." It can be necessary, but it is certainly not always evil. Zoning serves an important purpose in controlling and directing development. Any developer should be cautioned against building a park in an area without zoning; he or she must be reminded that anything might be built next door. There are other situations that call for careful judgment. For example, developers are well advised to resist the temptation to build in undesirable locations that can be rezoned easily.

The 1930s image of mobile homes as inexpensive housing for migrants and transients resulted in many communities passing restrictive zoning ordinances that pushed the development of mobile home parks outside of town or city borders. Although times are rapidly changing, there are many areas of the country where these restrictions are still enforced. The key to successfully challenging them may be (1) ensuring the protection of the property values of those who are envisioned as the permanent (i.e., voting) homeowning population and (2) demonstrating that mobile home park residents will not be a drain on local services and facilities (the latter is more difficult when state law dictates that new park residents will not be paying property taxes on their homes). Moreover, when a developer has achieved these two goals, he or she is more likely to prove that the proposed property will in fact be an asset to the community.

Approval may not be inexpensive. In at least a dozen states, for example, developers and property managers have had to fight for legislation at the state level to override local zoning ordinances.

Getting Help

A mobile home park manager or developer should be aware of all zoning regulations before proceeding with any type of construction project. Because zoning, and especially rezoning, can be a highly specialized subject, one should consider the services of individuals or companies who have had experience in this field. Such a consulting "team" can be invaluable in ascertaining little-known requirements and generating public support and political approval for zoning in favor of the park.

A team of consultants could consist of any or all of the following.

- Designer or architect
- Attorney
- Accountant
- Real estate appraiser
- Engineer
- Market feasibility expert
- Property manager
- Construction manager
- Public relations expert
- Government relations expert

The General Rezoning Procedure

Three basic meetings or presentations are necessary to meet the requirements of most rezoning procedures:

1. Preliminary meeting with the staff of the local planning and zoning commission. This is the most important of the three as it provides the opportunity to "test the waters." The objectives of this meeting should be: (a) to discern the commission members' reactions to the plans, (b) to determine the general compliance of the proposed project with current zoning ordinances and regulations, (c) to understand objections to the proposal, if any, and (d) to get ideas and recommendations that can be incorporated into a final plan.
2. Formal presentation before the planning and zoning commission. This will result in either approval or denial of the project. (An appeal process is usually available in the case of denial.)
3. Formal presentation before the city council or county commissioner.

Prehearing Meetings

Meetings with members of the planning commission, representatives of governmental agencies, and other city officials that occur before plans are finalized are not only cost-effective over the long term, but establish credibility. After all, "bureaucracies" consist of human beings who are more willing to help people they know to be taking the community's concerns into account. In the course of obtaining approval for a project — and before the preliminary meeting with the planning and zoning commissioners — meetings with specific individuals should be considered. These may include the city manager (or the highest ranking official responsible for development in the geographic area), the city or county engineer, the county assessor, the police chief (or ranking local police official), the fire chief (or ranking local fire official), utility departments heads, and the post office supervisor for the project location.

The park developer has specific objectives for these meetings. He or she should acquire a sense of what outsiders think of the plans, invite recommendations, and sell himself or herself as the person responsible for the project. None of this will come to anything, however, if the developer does not establish and maintain good community relations. The people who live near a proposed site for a new mobile home park must be in favor of the park (or at least not against it) if the rezoning request is to be successful. Citizen opposition to mobile home park development generally is based on historical misconceptions about mobile home parks and the people who live in them. A positive community viewpoint is generated by explaining plans, listening to others, and creating a quality project that reflects a response to the expressed local concerns. Experts in community, public, and government relations can be invaluable in researching people's attitudes, assessing local political realities, and developing necessary strategies.

In any negotiations with community groups, the park representative should make only those commitments that he or she intends to keep and preferably secure any agreements or approvals (in writing with signatures) before going to any formal meetings with officials.

A Sample Zoning Application Outline   A rezoning proposal to be presented before a local planning and zoning commission, city council, or county commissioner generally includes at least the following items.

- The formal rezoning application (official form)
- A narrative project description (including a description of the project site; a description of neighboring developments; a requested rezoning action with a discussion of proposed density, setbacks, open space, architectural compatibility, parking, and traffic at the site; and neighborhood benefits)
- Site plans, artist's rendering, and photographs

- A letter from the owners authorizing the individual who is representing them
- The assessor's quarter-section map, an enlarged subdivision map, and the subject parcel survey with legal descriptions
- A hydrology report
- Impact studies, as necessary (topics for analysis may include the environment, traffic, community services, economy, taxation, and/or schools)
- Documentation of neighborhood contacts
- Other documents required locally

Formal Presentations

Presentations for rezoning revolve around four major areas of local concern — general information, health considerations, community impact issues, and safety.

General Information   This may include several maps — one showing the relationship of the property to adjacent areas, roads, and the whole community; another displaying the existing zoning of the proposed project; and a site map (an overall concept plan which is usually prepared in color, often to a scale of one inch equals 100 feet). The developer should present landscaping plans as well as detailed designs showing site dimensions, home placement, offsets, amenities, open spaces, recreation areas, and entrances. He or she should have photographs of the property and the surrounding area, architectural renderings of buildings and signage, and an overall concept plan (which is usually prepared in color). Also included are preliminary engineering plans (not the final construction and engineering drawings), usually drawn to a scale of one inch equals 100 feet. Finally, the developer should provide examples of typical mobile homes and mobile home prices and explain how the mobile home park will be an asset to the community.

Health Considerations   Schematic drawings of systems (infrastructure) for the entire park — water distribution, storm and sanitary drainage, sewage collection and discharge outlets, and solid waste disposal are important inclusions to show how health considerations will be addressed. In all cases, the current zoning requirements should be indicated along with a comparison with the standards established for the project.

Community Impact   This information should be given to commission or board members prior to the formal public meeting because attorneys and accountants for the local government must have time to substantiate the documentation. The presentation should include a demographic profile of typical mobile home park residents (age, income, family size, etc.). Various impact studies may also be presented (e.g., an economic impact study to demonstrate the effect of the proposed park on surrounding property values). Other possible studies may analyze the park's anticipated impact on traffic patterns, the environment, community services, schools, etc. The developer can explain credit practices and CC&Rs (covenants, conditions, and restrictions, as for a homeowners' association) that will be used to screen prospective residents.

Safety   Always an important consideration, concerns about safety may be addressed by the development and presentation of a map. Items to be illustrated would include traffic flow, street light positioning, and patterns of ingress and egress for residents, visitors, those providing park services, and emergency vehicles. Security issues may also be covered.

Overcoming Objections and Denials

In the event a proposal for rezoning is denied, the park representative should be prepared to offer rebuttal — but only for those points on which the denial was based. It does not pay to argue in the negative. The park representative has to be prepared to make concessions and to give commission or council members something to approve. This is another area where professional advice, including that of an attorney, may help smooth the process.

A Final Word

The purpose of this entire chapter has been to touch upon some of the important legal issues associated with mobile home parks and their management. The material here is by no means comprehensive. As a result, one piece of advice warrants repeating: Every mobile home park manager should be aware of applicable federal, state, and local laws, and when questions arise, he or she should always seek professional counsel.

Subdivision Checklist

Appendix 3 is a copy of the checklists used by the Town of Ithaca, Tompkins County, New York State for realty subdivision approval.

<u>ARTICLE VI, SECTION 36.</u>
<u>PRELIMINARY SUBDIVISION PLAT CHECKLIST</u>

PROJECT NAME   _____

PROJECT NUMBER   _____

PREPARER   _____

  √   = ITEM SUBMITTED
  W   = WAIVED
 N/A = NOT APPLICABLE
COND = CONDITION OF APPROVAL

1. _____ Completed and signed Development Review Application. Development Review Escrow Agreement and Back-up Withholding Form (if required).(Only 1 copy each.)

2. _____ Payment of review fees.
Deposit of escrow.

3. _____ Fully completed and signed Short Environmental Assessment Form, Part I (SEAF), or Long Environmental Assessment Form, Part I (LEAF). (See Town Planner as to which to submit.)

4. _____ Estimated site improvement cost (excluding cost of land acquisition and professional fees) to be prepared (preferably) by a
licensed professional engineer.

5. _____ A preliminary plat with the following information must be filed
in the office of the Town Planner at least thirty (30) calendar
days prior to the Planning Board meeting at which preliminary
approval is requested.

a. _____ Vicinity Map showing the general location of the property, 1" =
1000' or 1" = 2000'.

b. _____ General layout, including lot lines with dimensions; block and
lot numbers; highway and alley lines, with 60-foot wide highway rights-of-way; areas to be reserved for use in common by
residents of the subdivision; sites for non-residential, non-pub-
lic uses; easements for utilities, drainage, preservation of scenic
views, or other purposes; and building setback lines, with
dimensions.

c. _____ General layout of the proposed highways, blocks, and lots
within the proposed subdivision. Tentative highway names.

d. _____ Contour intervals, to United States Geological Survey (USGS)
data, of not more than two feet when the slope is less than four
percent and not more than five feet when slope is greater than
four percent.

e. _____ Cultural features within and immediately adjacent to the proposed subdivision, including platted lots, highway improvements, bridges, culverts, utility lines, pipelines, power
transmission lines, other significant structures.

f. _____ Other significant structures within and immediately adjacent
to the proposed subdivision, including parks, wetlands, critical
environmental areas, and other significant features.

g. _____ Direction of flow of all water courses. Calculation of drainage
area above point of entry for each water course entering or
abutting the tract.

h. _____ Location and description of all section line corners and government survey monuments in or near the subdivision, to at

least one of which the subdivision shall be referenced by true courses and distances.

i. _____ Location, name, and dimensions of each existing highway and alley and each utility, drainage, or similar easement within, abutting, or in the immediate vicinity of the proposed subdivision.

j. _____ Natural features within and immediately adjacent to the proposed subdivision, including drainage channels, bodies of water, wooded areas, and other significant features. Identification of areas subject to flooding as indicated on HUD Flood Boundary Maps, Wetlands Maps.

k. _____ Width at building line of lots located on a curve or having non-parallel side lines, when required by the Planning Board.

l. _____ Names and addresses of owners of all parcels abutting the proposed subdivision.

m. _____ Names of recorded subdivisions abutting the proposed subdivision.

n. _____ Restrictive covenants, if any.

o. _____ Key map, when more that one sheet is required to present plat.

p. _____ Name of subdivision, which shall not duplicate the name of any other subdivision in the county.

q. _____ Name of planner, architect, engineer, land surveyor, landscape architect, or other person who prepared the sketch plat or preliminary plat.

r. _____ Name(s) and address(es) of the owner(s).

s. _____ Name(s) and address(es) of the subdivider(s), if the subdivider(s) is(are) not the owner(s).

t. _____ Map scale, in bar form, (1" = 50' or 1" = 100') and north point.

u. _____ Date of plat, and any applicable revision dates.

v. _____ Names of town, county, and state.

w. _____ Border lines bounding the sheet, one inch from the left edge and one-half inch from each of the other edges; all information, including all plat lines, lettering, signatures and seals, shall be within the border lines.

x. _____ Four dark-line prints of the proposed plat and 25 reduced copies of all sheets of the proposed site plan (no larger than 11" × 17") and copy of all other items required above (except Development Review Application and escrow forms).

y. _____ Four dark-line prints of improvement plans and information, if improvements are required.

PRELIMINARY SUBDIVISION CHECKLIST

IMPROVEMENT PLANS AND RELATED INFORMATION

6. Where improvements are required for a proposed subdivision, the following documents shall be submitted to the Planning Board:

a. _____ Detailed construction plans and specifications for water lines, including locations and descriptions of mains, valves, hydrants, appurtenances, etc.

b. _____ Detailed construction plans, profiles, and specifications for sanitary sewers and storm drainage facilities, including locations and descriptions of pipes, manholes, lift stations, and other facilities.

c. _____ Highway paving plans and specification.

d. _____ The estimated cost of:

1. _____ Grading and filling,

2. _____ Culverts, swales and other storm drainage facilities,

3. _____ Sanitary sewers,

4. _____ Water lines, valves and fire hydrants,

5. _____ Paving, curbs, gutters and sidewalks,

6. _____ Any other improvements required by *Town of Ithaca Subdivision* Regulations.

7. _____ The plan and profile of each proposed highway in the subdivision, with grade indicated, drawn to a scale of *1" = 50'* horizontal, and *1" = 5'* vertical, on standard plan and profile sheets. Profiles shall show accurately the profile of the highway or alley along the highway center line and location of the sidewalks, if any.

ARTICLE VI, SECTION 37. FINAL SUBDIVISION PLAT CHECKLIST

PROJECT NAME   _____

PROJECT NUMBER   _____

PREPARER   _____

  √   = ITEM SUBMITTED

  W   = WAIVED

 N/A = NOT APPLICABLE

COND = CONDITION OF APPROVAL

1. _____ Completed and signed Development Review Application.

    _____ Development Review Escrow Agreement and Back-up With-holding Form (if required). (Only 1 copy each.)

2. _____ Payment of review fees.

    _____ Deposit of escrow.

3. _____ Fully completed and signed Short Environmental Assessment Form, Part I (SEAF), or Long Environmental Assessment Form, Part I (LEAF). (See Town Planner as to which to submit.)

4. _____ Owner's Certificate: A certificate signed by the owner(s) to the effect that he/they own the land, that he caused the land to be surveyed and divided, and that he makes the dedications indicated on the plat.

5. _____ Surveyor's Certificate: A certificate signed and sealed by a registered land surveyor to the effect that (1) the plat represents a survey made by him, (2) the plat is a correct representation of all exterior boundaries of the land surveyed and the subdivision of it, (3) all monuments indicated on the plat actually exist and their location, size and material are correctly shown, and (4) the requirements of these regulations and New York State laws relating to subdividing and surveying have been complied with.

6. _____ Mortgagor's Certificate: A certificate signed and sealed by the mortgagor(s), if any, to the effect that he consents to the plat and the dedications and restrictions shown on or referred to on the plat.

7. _____ Two copies of the County Health Department approval of the water supply and/or sewage system.

8. _____ Certification signed by the Chairman or other designated official or agent of the Planning Board to the effect that the plat was given final approval by the Planning Board.

9. _____ A final plat must be filed in the office of the Town Planner at least thirty (30) calendar days prior to the Planning Board meeting at which final approval is requested. Final plat to include the following:

_____ Highway and alley boundary or right-of-way lines, showing boundary, right-of-way or easement width and any other information needed for locating such lines; purposes of easements.

_____ Highway center lines, showing angle of deflection, angles of intersection, radii, lengths of tangents and arcs, and degree of curvature, with basis of curve data. Lengths and distances shall be to the nearest one hundredth foot. Angles shall be to the nearest half minute.

_____ Highway names.

_____ Location, name, and dimensions of each existing highway and alley and each utility, drainage, or similar easement within, abutting, or in the immediate vicinity of the proposed subdivision.

_____ Exact boundary lines of the tract, indicated by a heavy line, giving the dimensions to the nearest one hundredth foot, angles to the nearest one-half minute, and at least one bearing; the traverse shall be balanced and closed with an error of closure not to exceed one to two thousand; the type of closure shall be noted.

_____ Location and description of all section line corners and government survey monuments in or near the subdivision, to at least one of which the subdivision shall be referenced by true courses and distances.

_____ Location of property by legal description, including areas in acres or square feet. Source of title, including deed record book and page numbers.

_____ Name and address of all owners of the property and name and address of all persons who have an interest in the property, such as easements or rights-of-way.

_____ Name(s) and address(es) of the subdivider(s), if the subdivider(s) is (are) not the owner(s).

_____ Accurate locations and descriptions of all subdivision monuments.

_____ Accurate outlines and descriptions of any areas to be dedicated or reserved for public use or acquisition, with the purposes indicated thereon; any areas to be reserved by deed covenant for common uses of all property owners in the subdivision.

_____ Building setback lines with dimensions.

_____ Lot lines, fully dimensioned, with lengths to the nearest one-hundredth foot and angles or bearings to the nearest one-half minute.

_____ Width at building line of lots located on a curve or having non-parallel side lines, when required by the Planning Board.

_____ Names and addresses of owners of all parcels abutting the proposed subdivision.

_____ Names of recorded subdivisions abutting the proposed subdivision.

_____ The blocks are numbered consecutively throughout the subdivision and the lots are numbered consecutively throughout each block.

_____ Key map, when more than one sheet is required to present plat.

_____ Vicinity map showing the general location of the property, 1" = 1000' or 1" = 2000'.

_____ Name of subdivision, which shall not duplicate the name of any other subdivision in the county.

_____ Name and seal of the registered land surveyor or engineer who prepared the topographic information. Date of survey.

_____ Name and seal of registered land surveyor who made the boundary survey. Date of survey.

_____ Name(s) and address(es) of the owner(s)

_____ Map scale (1" = 50' or 1" = 100') in *Bar Form* and north point.

_____ Date of plat and any applicable revision dates.

_____ Name of town, county, and state.

_____ Border lines bounding the sheet, one inch from the left edge and one-half inch from each of the other edges; all information, including all plat lines, lettering, signatures, and seals, shall be within the border lines.

_____ Reference on the plat to any separate instruments, including restrictive covenants, which directly affect the land in the subdivision.

_____ One original or mylar copy of the plat to be recorded and three dark-line prints, on one or more sheets.

ARTICLE VI, SECTION 38.
IMPROVEMENT PLANS AND RELATED INFORMATION

___√___ = ITEM SUBMITTED
___W___ = WAIVED
___N/A___ = NOT APPLICABLE
COND = CONDITION OF APPROVAL

1. Where improvements are required for a proposed subdivision, the following documents shall be submitted to the Planning Department:

_____ Detailed construction plans and specifications for water lines, including locations and descriptions of mains, valves, hydrants, appurtenances, etc.

_____ Detailed construction plans, profiles, and specifications for sanitary sewers and storm drainage facilities, including locations and descriptions of pipes, manholes, lift stations, and other facilities.

_____ Highway paving plans and specification.

_____ The estimated cost of:

a. _____ Grading and filling,

b. _____ Culverts, swales and other storm drainage facilities,

c. _____ Sanitary sewers,

d. _____ Water lines, valves and fire hydrants,

e. _____ Paving, curbs, gutters and sidewalks,

f. _____ Any other improvements required by Town of Ithaca Subdivision Regulations.

_____ The plan and profile of each proposed highway in the subdivision, with grade indicated, drawn to a scale of 1" = 50' horizontal, and 1" = 5' vertical, on standard plan and profile sheets. Profiles shall show accurately the profile of the highway or alley along the highway center line and location of the sidewalks, if any.

APPENDIX 4

Sample Covenants for a Realty Subdivision

DECLARATION OF RESTRICTIONS

DECLARATION OF RESTRICTIONS, by Walter J. Jones and Joyce Y. Jones, hereinafter designated as "Jones",

WHEREAS, Jones is the owner and developer of a certain tract of land known as "Historic Valley Estates", located on Scenic Road in the Town of Ithaca, Tompkins County, New York, more particularly described in Schedule A which is annexed hereto and made a part hereof, and

WHEREAS, said Jones is expecting to sell residential lots within said tract to individual homeowners and desires to subject the land and purchases thereof to certain restrictions, conditions and covenants for the purpose of maintaining the value and atmosphere desired for the subdivision,

NOW THEREFORE, Jones hereby declares that all lots shown on the tract of land described in Schedule A are held and shall be conveyed subject to the following restrictions, conditions and covenants:

1. All dwellings and accessory strictures shall conform to the building code of the State of New York and the zoning law of the Town of Ithaca.
2. **USE OF LOTS IN DEVELOPMENT:** All lots in the residential subdivision known as Historic Valley Estates shall be used primarily for residential purposes. However, an attached apartment or office of owner in residence not exceeding 40% of the building space shall be permitted.
3. Each lot may contain only one residence which shall not exceed three levels.

4. Refuse shall be removed no less often than once per week and shall be stored in locked sheds between collections.

5. All dwellings constructed shall have at least a partial basement which shall not be considered in counting "levels" as above described.

6. All accessory structures shall be located at least 10.0 feet from the street line and shall be of a style identical to the style of the house and the exterior finish of the accessory structure shall match that of the house constructed on that lot.

7. Fences may be erected only behind the dwelling located on any lot and shall be constructed only of wood or wrought iron except that decorative wrought iron fences not to exceed three feet in height may be installed on any portion of the lot.

8. Only one domesticated cat and one domesticated dog may be harbored on or about any lot. Dogs shall be leashed or penned at all times when allowed outside of a dwelling. However, no dog shall be allowed to bark in a manner which is offensive to any adjoining landowner.

9. No signs of any kind shall be displayed for public view on any lot except that a five square foot sign advertising a property for sale or rent may be employed when the property is being offered for conveyance.

10. No unlicensed motor vehicles may be placed on any premises unless they are kept in an enclosed building.

11. Any sound produced by other than natural phenomenon in excess of 60 decibels when measured at the property line nearest to the source shall be prohibited. This restriction does not apply to normal noise created by necessary maintenance of the premises.

12. No exterior antennae or satellite dish exceeding four feet in diameter shall be maintained on any lot.

13. DESIGN COMMITTEE: An Historic Valley Estates design committee shall be created to encourage design excellence, to foster careful design so that there is harmony, between the residences and their natural sights and among the residences themselves so as to preserve the natural beauty of the area. The committee shall consist of three members. The following persons are designated as the initial members of the committee: Walter J. Jones, Joyce Y. Jones and Thomas Korn. Walter J. Jones and Joyce Y. Jones shall be members for a period of five years. Thomas Korn shall be a member for one year. After completion of their designated terms, all members of the design committee shall be selected by a majority, of the 67 lot owners of Historic Valley Estates. Each lot shall be entitled to one vote. At least two members of the design committee must be Historic Valley Estates owners and the third member may be a professional architect or planner.

14. It shall be the duty of the design committee to adopt, amend and repeal rules and regulations to be known as the design committee rules and restrictions which interpret or implement the provisions of these covenants and restrictions. However, any such rules, regulations and restrictions promulgated by the design committee may be amended or repealed by a two-third (2/3) majority of the 67 lot owners of Historic Valley Estates. Each lot shall be entitled to one vote.

15. Invalidation of any one of these covenants, conditions and restrictions whether by judgment or court order shall not affect any of the remaining provisions which shall remain in full force and effect.

16. All restrictions, conditions and covenants herein shall run with the land and continue as such for twenty years from the date hereof. They shall be extended from that time automatically for successive periods of twenty years unless 90% of the owners of the 67 lots shown in Schedule A at that time shall agree to alter, modify or eliminate any or all of these restrictions.

17. The owner of Lot 1 consisting of .820 acres as shown on the above described Historic Valley Estates plat plan shall be subject to the following restrictions:

    (a) The easternmost portion of said land described on the said plat plan as a 50 foot "forever wild buffer" located adjacent to the west line of Scenic Road shall remain forever wild. No building or other structure may be placed thereon; however, the owners of Lot No. 1 shall maintain said area by removing dead trees from the area and may plant trees, shrubs and flowers in said area in their sole discretion.

18. A standard sewer line easement is hereby granted to the Town of Ithaca as shown on the Historic Valley Estate plat plan dated May 21, 1996.

APPENDIX 5

Appendix 5 is a copy of the "Mobile Home Park Regulation and Licensing Local Law of the Town of Newfield", Tompkins County, New York State. It was first adopted in 1989 as a mobile home park regulation for homes that we now recognize as manufactured homes.

Adopted July 26, 1989

Municipal Manufactured Home Community Code

LOCAL LAW NO. 2 FOR THE YEAR 1989

BE IT ENACTED BY THE TOWN BOARD OF THE TOWN OF NEW-FIELD AS FOLLOWS:

SECTION 1.0

Title, Statutory Authorization and Purposes

1.1 TITLE

This local law shall be known as the "Mobile Home Park Regulation and Licensing Local Law of the Town of Newfield"

1.2 STATUTORY AUTHORIZATION

This local law is adopted under the authority of Section 130, Subdivision 6, of the Town Law of the State of New York.

1.3 PURPOSE

It is the purpose of this local law to promote the health, safety, and general welfare of the residents of the Town of Newfield, by the proper regulation and licensing of mobile home parks to provide for a clean, safe, healthy and wholesome environment and living conditions within mobile home parks for the residents thereof.

SECTION 2.0

Definitions

2.1 DEFINITIONS

As user in this local law:

(A) PERSON shall mean an individual, association, partnership or corporation or any combination thereof and the agent or employee thereof.

(B) MOBILE HOME shall mean a detached, single family dwelling unit with any or all of the following characteristics:

(1) Manufactured as a relocatable dwelling unit intended for year round occupancy and for installation on a site without a basement;

(2) Designed to be transported, after manufacture on its own chassis and connected to utilities after placement on a mobile home stand;

(3) Designed to be installed as a single-wide or double-wide unit, with only incidental unpacking and assembling operations

(4) Designed and manufactured as the type of unit which would require, if built after January 15, 1974, a seal as provided for in the State Code for Construction and Installation of Mobile Home regardless of the actual date of construction.

(C) MOBILE HOME PARK shall mean any parcel of land or contiguous parcels of land under common ownership, containing three
(3) or more mobile homes, whether or not such mobile homes
are owned by the occupants thereof.

(D) MOBILE HOME LOT shall mean an area of land, in a mobile
home park, rented for the placement of a single mobile home
and any accessory structures incident thereto.

(E) LOT DEPTH shall mean the distance, measured along the cen-
terline of the lot, between the right-of-way line of a public street
or the pavement line of a private street, and the rear lot line.

(F) LOT WIDTH shall mean the distance between the two side lot
lines when measured perpendicular to the center-line of the lot.

(G) MOBILE HOME STAND shall mean that part of a mobile home
lot on which the mobile home is placed and which is constructed
in accordance with the standards provided in this local law.

(H) PARK OCCUPANT shall mean a person or persons living in a
mobile home in a mobile home park.

(I) PARK OPERATOR shall mean the person or persons owning a
mobile home park and/or responsible for on-site management
and operation of a mobile home park.

(J) SITE PLAN shall mean a drawing(s) submitted to the Town Clerk
as part of the application for a license for a mobile home park
and containing all the information required by this local law in
sufficient detail to enable the required reviews.

SECTION 3.0

License Required

3.1 MOBILE HOME PARK

No person or persons, being the owner or occupant of any land in the
Town of Newfield shall use or permit the development and use of such land
as a mobile home park without first obtaining a license therefor as provided
in Section 4.0 of this local law. Such license shall be renewed every two
(2) years.

SECTION 4.0

Application for a License

4.1   APPLICATION FOR A LICENSE

Written application for a license for a mobile home park shall be filed with the Town Clerk of the Town of Newfield upon forms provided for such purpose along with the requisite fee. The Clerk shall submit said application to the Town Board or its designee for review and determination.

4.2.   CONTENTS OF APPLICATION

Applications for a mobile home park license shall include, but not be limited to, the following:

(A) Applicant: Names and addresses of all applicants, if an individual or partnership, and the name and address of principal officers and shareholders if applicant is a corporation;

(B) Land Owner: Name and address of the owner of land upon which the mobile home park is to be located if other than the applicant;

(C) Map: Location map;

(D) Design and Layout: Scaled sketch drawings of the proposed mobile home park indicating its design and layout and demonstrating conformity with the requirements of Sections 5.0 and 6.0 of this local law;

(E) Water and Sewer: If public water and/or sewage systems are not to be used, application for approval of the proposed mobile home park by the Tompkins County Health Department must be submitted with the application;

(F) Buildings, Parking, Open Areas: Scaled sketch, plans or written descriptions of all buildings, streets, parking areas, recreation and open spaces, and landscaping to be constructed or provided within the mobile home park;

(G) Topography and Drainage: An indication of existing topography and drainage patterns including wet or swampy areas;

(H) Rules and Regulations: A copy of all contemplated park rules, regulations and covenants; a list of management and tenant responsibilities; a written statement of any entrance and existing fees, if any, utility connection fees, if any, and any security deposits to be charged, if any;

    (1) Additional Information: Such further information as the developer may feel is necessary to describe the intent and ability to comply with the environmental, health, and safety standards of this local law.

4.3   PROCEDURE

The application for a mobile home park license shall be filed in duplicate with the Town Clerk along with the required fee and the following procedure shall apply:

(A) Application Review: The enforcement officer shall refer one copy of the application to the Town Board for review of the layout and design of the proposed park. If the application shall be complete, it shall be referred to the Town Board for further action. Incomplete applications shall be rejected.

(B) Public Hearing: Within sixty (60) days from receipt of the application the Town Board shall hold a public hearing on said application which hearing shall be duly advertised on ten (10) days advance notice to the public. Any interested party may speak at the hearing.

(C) Approval/Disapproval: Within forty-five (45) days from the date of the public hearing the Town Board shall approve or disapprove the application, and set forth any special conditions as may be required, and instruct the Town Clerk to issue a temporary permit based upon such approval and upon issuance of the building permit required under the Now York State Uniform Fire Protection and Building Code. Issuance of a temporary permit is authorization for the applicant to proceed with the final plans for the mobile home park incorporating the conditions attached to said temporary permit.

(D) Final Plan: Final plans for the proposed mobile home park, or, if construction is to be staged, that portion of it to be constructed initially, shall be submitted to the Town Board for review within one (1) year from the date of issuance of the temporary permit. If such submission is not made, the temporary permit shall be withdrawn unless extended by the Town Board for good cause shown.

(E) Conditions: The Town Board shall determine if the conditions imposed have been met and shall be concerned with such things as the appropriateness and quality of the overall site plan in terms of the effective use of the site, suitability of proposed landscaping, usefulness of proposed recreation areas, and the general visual character of the park. In addition, the Town Board shall determine that the plans comply with the requirements of Sections 5.0 and 6.0 of this local law.

(F) Issuance of License: Within forty-five (45) days from the receipt of final plans the Town Board shall approve the final plan and instruct the Town Clerk to issue a license for the mobile home park. Final plans may in the Town Board's discretion be conditionally approved or disapproved. If disapproved, the temporary permit may be canceled or extended for good cause shown at the option of the Town Board.

4.4   RENEWAL OF MOBILE HOME PARK LICENSE

The Town Board shall renew a mobile home park license every two (2) years from the date of issuance. If the mobile home park has not been constructed in accordance with the approved plans and all conditions attached hereto, or if a violation of this local law shall be found, or if any unapproved change shall take place, the license shall not be renewed until said mobile home park has been brought into compliance. In such case, the Town Board shall serve an order upon the holder of the license in accordance with the provisions of Section 10 of this local law.

4.5   LICENSES FOR EXISTING MOBILE HOME PARKS

The owner of any mobile home park existing prior to the adoption of this local law shall apply for a mobile home park license within sixty (60) days from the date of adoption of this local law and such license shall be subject to renewal after one (1) year on the initial license and is to be renewed every two (2) years thereafter. Upon initial application, the Town Board shall issue a temporary license valid for one (1) year and shall serve notice on the park owner of any violations of this local law which might exist or any improvements necessary to meet the requirements of this local law.

No license for a mobile home park existing at the time of enactment of this local law shall be renewed until violations cited by the Town Board have been corrected and a renewal has been authorized.

4.6   FEES

Each application shall be accompanied with a fee of one hundred dollars ($100.00) for the first ten (10) mobile home units plus five dollars ($5.00) for each additional mobile home unit, which fee shall be retained by the Town of Newfield regardless of what disposition is made of the application; such application fees shall be payable to the Town Clerk.

4.7   ADDITIONS TO LICENSED MOBILE HOME PARKS

Any addition of new mobile home lots, to any mobile home park licensed herein or operating as of the effective date of this law, shall be subject to approval in the same manner as a new mobile home park and the requirements Bond conditions pertaining to parks established after the effective date of this local law shall be applicable to such new lots or additions or expansions. A license to operate a mobile home park shall not confer upon the holder any rights to expand the number of units, nor alter any approved plan without the approval of the Town Board under the procedures set forth herein.

4.8   LICENSES NON-TRANSFERABLE

No license issued under this local law shall be transferable. It shall be deemed a transfer if any corporate licensee shall transfer more than fifty percent (50%) of its stock to parties not shareholders at the time of the issuance of the corporate licensee's license.

SECTION 5.0

Environmental Requirements

5.1   COMPLIANCE WITH APPLICABLE LAWS

All applicants shall comply with the provisions of Article 8 of the Environmental Conservation Law, including, but not limited to, the requirements of Section 7.2 thereof.

5.2   SITE LOCATION

(A) Neighborhood Facilities: Mobile home park plans shall include a statement concerning the availability of shopping facilities and fire protection services in relation to the location of the proposed park. A statement from the appropriate school district official shall also be included indicating that school bus service will be provided, if necessary, and evaluating the impact of the park on the school system.

(B) Relationship to Major Roads: Mobile home park plans shall include a sketch of the site as it relates to major traffic arteries with indications of anticipated traffic patterns to the park. Direct connections onto major highways shall be in accordance with the standards set forth in Section 6.6 of this article.

5.3   NATURAL FEATURES

(A) General Requirements: Topography, groundwater level, surface drainage, and soil conditions shall not be such as to create hazards to the property or to the health and safety of the occupants. No developed portions of the site shall be subject to excessive settling or erosion. A sloping site should be graded to produce terraced lots for placement of the mobile home units and, in general, units should be placed parallel rather than perpendicular to the slope.

(B) Surface Drainage: Mobile home park plans shall show all proposals for changes in existing surface drainage patterns. All parks shall be graded to prevent ponding of surface water. If any part of the site is located in a floodplain no structure of mobile home shall be located on land designated as a 100-year floodplain area as determined by the U.S. Corps of Engineers or other official agencies unless they meet the specifications outlined by the Federal Government for development of a mobile home park in a flood plain as shall be in effect at the time of application and shall not otherwise be prohibited by federal or state law.

(C) Soils: Soils should be of sufficient bearing and stability properties to provide adequate support for mobile home installations. Topsoil should be of sufficient depth to sustain lawns, trees, and other vegetation.

(D) Natural Features: Mobile home park plans shall show existing tree masses or trees over six (6) inches in diameter at breast height, hedgerows, and other notable existing natural features such as streams or rock formations. Such natural features shall be retained as much as possible in the site plan and densities shall be reduced, if necessary, to permit such retention.

5.4   LOT LAYOUT AND UNIT PLACEMENT (Applicable only to lots installed after effective date of the local law.)

(A) Overall Considerations

(1) Required Separation: Mobile home units may be positioned in a variety of ways within a park provided that a separation of at least thirty (30) feet is maintained between units. A drawing showing the proposed layout of mobile home units shall be prepared.

(2) Setback: No mobile home shall be located less than twenty-five (25) feet from the pavement edge of a private park street or fifteen (15) feet from the right-of-way of any public street within a mobile home park. A minimum of fifty (50) feet shall be maintained between a mobile home unit and any property line abutting a public road or highway.

(B) Density: The density of development in a mobile home park shall not exceed 4.0 units per gross acre.

(C) Minimum Lot Size: Mobile home lots shall be a minimum of 6,000 square feet in area and shall have a minimum width of 55 feet. In special cases, where unusual park design provides for wider streets or a greater amount of usable recreation or public open square than required by this local law, or when other special conditions exist, the Town Board may approve a modification of lot size. In no case, however, shall the gross density, as specified in (B) above be extended, nor shall the lot areas be reduced below 5,000 square feet, nor the lot width be reduced below 50 feet.

5.5   VEHICULAR CIRCULATION AND STORAGE

(A) Park Road layout: A drawing of the proposed park road layout, including connections to be made to adjacent existing roads or highways, shall be included in all mobile home park plans.

Straight, uniform gridiron road patterns should be avoided unless they can be relieved by mobile home clustering, landscaping, and an interesting open space system. Turn arounds shall be provided sufficient to handle all emergency and trash removal vehicles.

(B) Park Road Construction: Roads within a mobile home park shall be adequately paved and maintained at all times and shall be of sufficient width for applicable traffic within the mobile home park. Driveways for lots shall be delineated and marked.

(C) Off-Street Parking. A minimum of two (2) off-street parking spaces shall be provided for each mobile home site. Such spaces may be located on the individual lot or grouped to serve two (2) or more mobile home sites. Parking areas shall be adequately paved and maintained.

(D) Storage Space for Auxiliary Vehicles: Adequate storage space shall be provided for any travel trailers, camper, boat, snowmobile, or similar auxiliary vehicle or conveyance parked or stored on any mobile home lot. Off-street parking space required by Section 5.5(C) of this local law shall be used by passenger vehicles only and a supplemental parking area shall be provided in each park for the storage or temporary parking of all auxiliary vehicles. (Applicable only to lots installed after effective date of the local law.)

5.6   PARK ENTRANCE

(A) Entrance Roads: Each mobile home park shall provide for two (2) independent connections with existing public streets, such connections to be designed so that traffic can be maintained even though one access may be temporarily closed. A divided entrance road twenty five (25) feet in length providing at least ten (10) feet between entrance and exit lanes that are at least twenty (20) feet wide shall satisfy the requirements of this section.

(B) Sufficient Road Width: At points where traffic enters and leaves the park, road widths shall be sufficient to permit free and safe movement to or from the public street.

(C) Entrance Signs: Any sign located within a mobile home park shall comply with existing regulations and shall be located so as not to obstruct the visibility of motorists entering or leaving the park.

(D) Mobile Home Lot Adjacent to Park Entrance: No mobile home lot shall after the effective date of the local law be located less than fifty (50) feet from the intersection of a park entrance road and a public highway and no private mobile home driveway shall make a direct connection with an existing public highway.

5.7   MOBILE HOME SALES AREA

(A) Display and Sale: The display and sale of mobile homes shall not be permitted within any mobile home park unless they were in operation prior to the effective date of this local law. A reasonable number of "model" mobile homes, relative to the size of the park in general, may be set up temporarily within the park for display purposes, provided such operator shall also have a separate permanent display and sales area with separate parking facilities for customers located outside the park.

(B) Sales Area: In any area where mobile home sales are permitted such sales area shall be adequately paved and maintained with a hard base and shall be a dust-free surface and should contain a minimum of six (6) off-street parking spaces for customers. No display unit shall be located less than fifteen (15) feet from a public right-of-way.

5.8   COMMUNITY FACILITIES AND ACTIVITIES

(A) Plan Details: If community facilities and activities such as meeting rooms, recreation buildings, laundry rooms, and swimming pools are to be included in the mobile home park, the plan shall include details of these facilities and the owner's statement of intent to provide adequate supervision and management of such facilities and activities.

(B) Landscaping: All community facilities and activities shall be landscaped with trees, shrubs and grass and shall provide adequate paved off-street parking space.

(C) Location of Facilities: Community facilities and activities shall be located and designed in a manner that will be a visual asset to the mobile home park, and constructed of material that will be compatible with the residential character of the park.

5.9   OPEN TREATMENT AND PARK AMENITY

(A) Open Space and Developed Recreation Areas: In all mobile home parks a variety of open spaces shall be provided so as to be usable by and easily accessible to all park residents. Such open space shall be provided on the basis of 500 square feet for each mobile home unit with a total minimum requirement of 12,000 square feet. Part of all of such open space shall be in the form of developed recreation areas located in such a way, and of adequate size and shape, as to be usable for active recreation purposes. (This provision shall not apply to mobile home parks operating prior to the effective date of this local law.)

All open spaces shall be stabilized by grass or other forms of ground cover which will prevent dust and muddy areas.

(B) Buffer Zone: Mobile home parks located adjacent to residential, industrial or commercial development, or a heavily traveled highway, shall be buffered from such development or highway by a hedge or similar landscape screen which will rapidly reach a height of at least six (6) feet. A combination of landscaping and decorative fencing may be substituted provided the height requirement is met and considerable landscaping is used. (This provision shall not apply to parties operating prior to the date of this local law.)

(C) Soil Arid Ground Cover Requirements: Exposed ground surfaces in all parts of any mobile home park shall be paved, surfaced with crushed stone or other solid material, or protected with grass or plant material capable of preventing erosion and of eliminating objectionable dust.

(D) Trees: At least one tree shall be planted on each mobile home lot if no such tree already exists. Planted trees shall be a caliper of at least two (2) inches. (This provision shall not apply to parties operating prior to the date of this local law.)

(E) Walkways: Each mobile home stand shall be provided with a walkway leading from the stand to the street or to a driveway or parking area connecting to the street. Such walkway shall be adequately marked and maintained.

(F) Fencing: If fencing of individual lots within the mobile home park is to be provided by the mobile home occupant, standards shall be provided by the park operator so that consistency can be maintained.

(G) Park Lighting: All mobile home parks shall be furnished with adequate lights to provide sufficient illumination for the safe movement of vehicles and pedestrians at night over streets, driveways and walkways. Electric service to such lights shall be installed underground and decorative lighting fixtures shall be used where possible.

5.10 MOBILE HOME STAND

(A) Installation: Installation of mobile homes and the mobile home stand shall be made in accordance with the applicable provisions of the New York State Uniform Fire and Building Code as enforced by the Town of Newfield.

(B) Patios/Decks: Each mobile home site constructed after the effective date of this local law shall be provided with a patio or deck or combination of both with a minimum width of ten (10) feet and a total area of at least 200 square feet. Such patio or deck shall be constructed in accordance with adequate and usable materials and shall be properly maintained, and shall be located so that good access to the front door of the mobile home will be maintained.

(C) Accessory Buildings: No outdoor storage of personal property by mobile home tenants other than as provided in Section 5.5(D) shall be permitted unless the mobile home park operator shall provide or shall require each occupant to provide an accessory storage building. Such building shall not exceed 250 square feet in size, shall be a standard prefabricated product, and shall be installed on 6 poured concrete slab or other adequate foundation provided by the park operator. The location of the accessory building shall be determined by the park operator either at the time the park is developed or as sites are occupied. Accessory buildings shall be maintained.

5.11   HOME UNITS

(A) Unit Installation and Skirting: At the time of installation of the mobile home, the tires and wheels, and the hitch, if possible, shall be removed and the unit shall be secured, blocked, leveled, and connected to the required utility systems and support services. The mobile home shall be completely skirted within ninety (90) days of occupancy. Materials used for skirting shall provide a finished exterior appearance and shall be similar in character to the material used in the mobile home that would be permitted under the New York State Fire Prevention and Building Code for a crawl space in a conventional home. Skirting and/or a skirting system shall be installed and maintained in a professional manner.

(B) Expansions and Extensions: Expandable rooms and other extensions to a mobile home unit shall be supported on a stand constructed in accordance with construction standards for the mobile home stand. Skirting shall be required around the base of all such expansions or extensions. All expansions and extensions shall be built of such materials and designed in a fashion that the original mobile home and the expansion and extension shall appear to have been manufactured or constructed together as a single unit.

(C) Entrance Steps: Entrance steps shall be installed at all doors leading to the inside of the mobile home. Such steps shall be constructed of materials intended for permanence, weather resistance, and attractiveness and shall be equipped with handrails which will provide adequate support for users.

SECTION 6.0

Support Services and Utility Delivery Systems

6.1   WATER AND SEWER

(A) Public Water/Sewer: Mobile home parks hooked up to public water and/or sewer shall at all times be operated in accordance with the applicable rules and regulations of the water and/or sewer district.

(B) Private Water/Sewer: Mobile home parks using private water and/or sewage facilities shall at all times be operated in accordance with applicable laws, rules and regulations of the State of New York and Tompkins County Health Department.

6.2   SOLID WASTE DISPOSAL

(A) General: The storage, collection and disposal of solid waste in the mobile home park shall be so conducted as to create no health hazards, rodent harborage, insect breeding areas, accident or fire hazards or air pollution.

(B) Group Storage Areas: If group solid waste storage areas are provided for park occupants they shall be enclosed or otherwise screened from public view and shall be rodent and animal proof and located not more than 100 feet from any mobile home site they are to serve. Containers shall be provided in sufficient number to properly store all solid waste produced.

(C) Individual Storage Areas: Any solid waste containers stored on individual mobile home sites shall be screened from public view and shall be rodent and animal proof.

(D) Burning Prohibited: Disposal of solid waste by burning is expressly prohibited.

6.3   ELECTRIC POWER TELEPHONE AND TELEVISION SERVICE

(A) Electric: The mobile home park electrical distribution shall be installed underground and shall comply with the national electric code and with requirements of the utility company serving the area and the Public Service Commission.

(B) Telephone: The distribution system for telephone service shall be underground in accordance with the standards established by the New York Telephone Company.

(C) Television: Television service which is provided by a cable system shall be installed underground. When cable service is not available, a common antenna shall be provided with direct burial cable to each mobile home site.

6.4   FUEL SYSTEMS

All mobile home parks shall be provided with facilities for the safe storage of necessary fuels. All systems shall be installed and maintained in accordance with the applicable federal, state and local laws, codes and regulations governing such systems.

(A) Natural Gas: Natural gas installations shall be planned and installed so that all components and workmanship comply with the requirements of American Gas Association, Inc. and conform to the requirements, inspections and approval of the utility which will supply this product.

(B) Fuel Oil: Fuel oil systems with either common or individual supplies shall be designed, constructed, inspected and maintained in conformance with the provisions of National Fire Protection Association, Standard 30. All fuel oil storage tanks, whether provided as a bulk supply for a group of mobile homes or on each individual mobile home lot, shall be located and installed under applicable federal and state laws and regulations and shall be supplied with permanently installed and secured piping. Fuel oil tanks shall be located to the rear of the mobile home site and shall be landscaped and screened from public view.

(C) Liquified Petroleum Gas: Liquified petroleum gas systems shall be selected, installed and maintained in compliance with the requirements of National Fire Protection Association, Standard 58. LPG tanks shall be located to the rear of the mobile home site and shall be landscaped and screened from public view.

6.5   FIRE PROTECTION

(A) Fire District Rules: The mobile home park plan shall include a list of the applicable rules and regulations of the fire district wherein said park is located and shall comply with such rules and regulations.

(B) Hydrants: If the mobile home park is located in a public water district, fire hydrants shall be installed in accordance with the requirements of the district and inspected and approved by the designated local official.

(C) Safe Maintenance: Mobile home parks shall be kept free of litter, rubbish and all other flammable materials.

(D) Hydrant and Fuel Storage Map: The mobile home park operator shall furnish the Newfield Fire Department, Tompkins County Sheriff's Department and New York State Police with a map and plan of the mobile home park, which shall designate the location of all fire hydrants and fuel storage areas, if any.

SECTION 7.0

Mail service

7.1   MAILBOX PLACEMENT

(A) Location: Mailbox location shall provide safe and easy access for the pickup and delivery of mail.

(B) Cluster Delivery: Grouped mailboxes for cluster delivery shall be located in a way that will not require stopping on a public right-of-way for pickup.

(C) Landscaping: When mailboxes are grouped together for some form of cluster delivery such groupings shall be landscaped.

SECTION 8.0

Park Operations and Maintenance

8.1   RESTRICTIONS ON OCCUPANCY

(A) Length of Placement: In any mobile home park, no space shall be rented for the placement and use of a mobile home for residential purposes except for periods in excess of 180 days.

(B) Mobile Home Qualifications for Placement:

   (1) No mobile home manufactured after January 15, 1974 shall be admitted to any park after the effective date of this local law unless it bears the seal issued by the State of New York

and required by the State Code for Construction and Installation of Mobile Homes or has met applicable Federal HUD standards for its manufacture.

(2) No mobile home manufactured prior to January 15, 1974 shall be admitted to any park after the effective date of this local law unless it shall have been built to satisfactory standards and is still in serviceable condition. All such mobile homes shall be inspected by the Code Enforcement officer prior to their installation in any mobile home park.

(3) Notwithstanding subsections (B)(1) and (B)(2) herein, no mobile home, regardless of its date of manufacture, shall after the effective date of this local law, be admitted to any park if such mobile home has deteriorated or been damaged to the extent whereby it shall no longer be adequate for reasonable human habitation or shall have a deteriorated or damaged external appearance.

8.2   RESPONSIBILITIES OF PARK OPERATOR

(A) Compliance and Supervision by Operator: The person to whom a license for a mobile home park is issued shall operate the park in compliance with this local law and shall provide adequate supervision to maintain the park, its common grounds, streets, facilities and equipment in good repair and in a clean and sanitary condition.

(B) Compliance by Occupants: The park operator shall notify park occupants of all applicable provisions of this local law and inform them of their responsibilities and any regulations issued thereunder.

(C) Placement of Mobile Home: The park operator shall place or supervise the placement of each mobile home on its mobile home stand which includes ensuring its stability by securing and installing all utility connections.

(D) Register of Occupants: The park operator shall maintain a register containing the names of all occupants and the make, year and serial number, if any, of each mobile home. Such register shall be available by the park owner on a 24 hour emergency basis to police and fire department officials, and available during normal business hours to authorized persons inspecting the park and officials of the Town of Newfield.

8.3   RESPONSIBILITIES OF PARK OCCUPANTS, ENFORCEMENT BY PARK OPERATOR

(A) General: The park occupant shall be responsible for the compliance of any of the provisions of this local law within his or her control and ability.

(B) Maintenance of Mobile Home: The park occupant shall be responsible for the maintenance of his mobile home and any appurtenances thereto, and shall keep all yard space on his site in a neat and sanitary condition.

(C) Maintenance of Lot: It shall be the responsibility of each mobile home occupant to keep his site free of litter, rubbish, unused vehicles and equipment or parts thereof.

(D) Compliance by Park Owner: The park owner shall at all times be responsible for the compliance with the provisions of this local law, whether or not a particular mobile home park occupant is also responsible. It shall be a responsibility under this local law for the park operator to require and enforce compliance, to the extent permitted by law, of the requirements herein as it shall apply to their tenants.

(E) Park Owner's Right to File Complaint: A park owner shall have the right to file a complaint against any tenant for an applicable violation of this local law if, after reasonable effort, such park owner shall be unable to obtain compliance by such tenant. The filing of such complaint, however, shall not in and of itself relieve such park owner from his/her obligations as licensee under this local law.

SECTION 9.0

Inspection

9.1   ENFORCEMENT

This local law shall be enforced by the Town Board of the Town of Newfield through the code enforcement officer. Said officers and their inspectors shall be authorized and have the right in the performance of duties to enter any mobile home park and make such inspections as are necessary to determine satisfactory compliance with this local law and regulation issued hereunder. Such entrance and inspection shall in routine cases be accomplished at reasonable times, after prior notice to the park operator, and in cases involving violations or in emergencies whenever necessary. Owners, agents or operators of a mobile home park shall be responsible for providing access to all parts of the premises within their control to the enforcement officer or to his inspectors, acting in accordance with the provisions of this section.

9.2   INSPECTION

It shall be the duty of the code enforcement officer to make regular inspections of all licensed premises as he shall deem necessary and shall inspect each licensed premises no less than once every year and within 60 days prior to the date of renewal of any mobile home park license and to investigate all complaints made under this local law.

SECTION 10.0

Criminal Penalties and Enforcements

10.1 ENFORCEMENT OFFICER

The Town of Newfield shall have the authority to appoint an enforcement officer authorized and empowered to act on behalf of the Town of Newfield to enforce the provisions of this law, including the right of entry onto any licensed premises or premises which are unlicensed, but reasonably deemed to be in violation of the law. The enforcement officer shall have the authority to issue appearance tickets returnable in the Town Justice Court with respect to any violation herein without specific direction of the Town Board.

10.2 PENALTIES

(A) Violations: Any person including a park operator or owner and/or mobile home tenant who commits or permits the commission of any act or acts in violation of any of the provisions of this local law shall be subject to a fine of not more than Two Hundred Fifty Dollars ($250.00) or imprisonment for not more than fifteen (15) days, or both such fine and imprisonment, and/or suspension of the license for a period of at least five (5) days, for each such violation. Each day such violation shall continue or be permitted to exist shall constitute a separate violation as shall be permitted by law.

(B) Additional Proceedings: In addition to the penalties herein provided for, the Town Board may also maintain an action or proceeding in the name of the Town in a court of competent jurisdiction to compel compliance with or to restrain by injunction any violation of this local law.

(C) Correction by Town: Notwithstanding any other penalty herein the Town of Newfield on written notice thirty (30) days after a conviction under this local law may enter upon the premises of the violation and take such steps necessary to correct any violation if the Town of Newfield shall determine such steps are in the public interest and in the interest of the inhabitants of the mobile home park and charge the violator for the reasonable costs thereof. Such unpaid charges shall be deemed town charges and shall be levied as in a manner of a special assessment on the tax levy against such property at the first levy following the billing for such charges, by the Town of Newfield. Such unpaid assessment shall be a lien against the real property of the violator.

SECTION 11.0

Revocation of License

11.1 INITIAL ORDER

Upon determination by the Town Board that there has been a violation of any provisions of this local law, they shall in addition or in lieu of any other penalty set forth in Section 10.2, may serve upon the holder of the license for such mobile home park and initially order, in writing and by certified mail, return receipt, directing that the conditions therein specified be corrected within ninety (90) days after the date of delivery of such order. The order shall also contain an outline of remedial action which, if taken, will effect compliance.

11.2 NOTICES

If, after the expiration of such ninety (90) day period, such violations are not corrected, the Town Board shall serve a notice in writing upon such mobile home park operator, requiring the holder of the park license to appear before the Town Board of the Town of Newfield at a time to be specified in such notice, to show cause why the mobile home park license should not be revoked. Such hearing before the Town Board shall occur not more than forty-five (45) days after the date of service of said notice by the Town Board.

11.3 HEARING

Within ten (10) days after the hearing at which the testimony and witnesses of the Town Board and the mobile home park license holder shall be heard, the Town Board shall make a determination, in writing, sustaining, modifying, or withdrawing the order issued by said Town Board as directed by Section 11.1 of this local law. Failure to abide by any Town Board determination to sustain or modify the initial order of said Town Board, and to take corrective action accordingly, shall be cause for the revocation of the mobile home park license affected by such order and determination.

11.4 NOTICE TO HEALTH DEPARTMENT AND TENANTS

The Town Clerk shall promptly notify the Tompkins County Health Department and all tenants of the subject mobile home park of any revocation of a license.

SECTION 12.0

Variances

12.1 VARIANCE PERMITTED FOR HARDSHIP

Where there are practical difficulties, unusual circumstances, or design innovations involved, the Town Board may grant variances from any of the provisions and regulations of this local law except those related to Health Department, Department of Environmental Conservation, and building code requirements. There shall be no right to a variance, the issuance of which shall be solely within the discretion of the Town Board.

12.2 APPLICATION

Application for a variance shall be in writing from the person applying for the mobile home park license required in accordance with Section 4.0 of the local law. In considering a request for a variance the Town Board shall be guided by the circumstances of the situation and the intent of the applicant, and shall act as to protect the best interests of the community.

SECTION 13.0

Appeals

13.1 APPEAL FROM DECISION OF ENFORCEMENT OFFICER

Any person aggrieved by any decision of the enforcement officer may take an appeal to the Town Board except with respect to any case of violations pending before the Town Justice Court. Said Board shall act in accordance with the provisions of Section 10.0 of this local law.

13.2 APPEAL FROM DETERMINATION BY TOWN BOARD

Any determination made by the Town Board under this local law, excepting in the case of violations heard in the Town Justice Court, may be reviewed by the Supreme Court under Article 78 of the Civil Practice Law and Rules.

SECTION 14.0

Saving Clause and Effective Date

14.1 SAVING CLAUSE

If any section, paragraph, subdivision or provision of this local law shall be adjudged invalid or held unconstitutional, the same shall not affect the validity of this local law as a whole or any part or provision thereof other than the part so decided to be invalid or unconstitutional.

14.2 EFFECTIVE DATE

This law shall take effect upon its filing with the Secretary of State in accordance with Section 27 of the Municipal Home Rule Law.

Adopted August 23, 1989

LOCAL LAW No. 3 FOR THE YEAR 1989

BE IT ENACTED BY THE TOWN BOARD OF THE TOWN OF NEW-
FIELD AS FOLLOWS:

SECTION 1.0

Title, Statutory Authorization and Purposes

1.1   TITLE

This local law shall be known as the "Amendment to Mobile Home
Park Regulation Bond Licensing Local Law of the Town of Newfield."

1.2   STATUTORY AUTHORIZATION

This local law is adopted under the authority of Section 130, Subdivision 6, of the Town Law of the State of New York.

1.3   PURPOSE

It is the purpose of this local law to promote the health, safety, and
general welfare of the residents of the Town of Newfield, by the proper
regulation and licensing of mobile home parks to provide for a clean, safe,
healthy and wholesome environment and living conditions within mobile
home parks for the residents thereof.

SECTION 2.0

2.1   Section 5.4(C) of Local Law No. 2 for the Year 1989 shall henceforth
      read:

   (C) Minimum Lot Size: Mobile home lots shall be a minimum of 6,000
        square feet in area and shall have a minimum width of 55 feet. In
        special cases, where unusual park design provides for wider streets
        or a greater amount of usable recreation or public open square
        than required by this local law, or when other special conditions
        exist, the Town Board may approve a modification of lot size.

2.2   Section 5.5(B) of Local Law No. 2 for the Year 1989 shall henceforth
       read:

    (B) Park Road Construction: Roads within a mobile home park shall
       be adequately paved and maintained at all times and shall be of
       sufficient width for applicable traffic within the mobile home
       park. Driveways for lots shall be drained and maintained silt free.

2.3   Section 5.5(C) of Local Law No. 2 for the Year 1989 shall henceforth
       read:

    (C) Off-Street Parking: A minimum of two (2) off-street parking
       spaces shall be provided for each mobile home site. Such spaces
       may be located on the individual lot or grouped to serve two (2)
       or more mobile home sites. Parking areas shall be adequately
       drained and maintained silt free.

2.4   Section 5.9(E) of Local Law No. 2 for the Year 1989 shall henceforth
       read:

    (E) Walkways: Each mobile home stand shall be provided with a
       walkway leading from the stand to the street or to a driveway or
       parking area connecting to the street. Such walkway shall be
       adequately maintained.

SECTION 3.0

Saving Clause and Effective Date

3.1   SAVINGS CLAUSE

    If any section, paragraph, subdivision or provision of this local law
shall be adjudged invalid or held unconstitutional, the same shall not affect
the validity of this local law as a whole or any part or provision thereof other
than the part so decided to be invalid or unconstitutional.

3.2   EFFECTIVE DATE

    This law shall take effect upon its filing with the Secretary of State in
accordance with Section 27 of the Municipal Home Rule Law.

LOCAL LAW NO. 1 FOR THE YEAR 1991

BE IT ENACTED BY THE TOWN BOARD OF THE TOWN OF NEW-FIELD AS FOLLOWS:

SECTION 1.0

Title, Statutory Authorization And Purposes

1.1   TITLE

This local law shall be known as the "Second Amendment to Mobile Home Park Regulation and Licensing Local Law of the Town of Newfield.

1.2   STATUTORY AUTHORIZATION

This local law is adopted under the authority of Section 130, Subdivision 6, of the Town Law of the State of New York.

1.3   PURPOSE

It is the purpose of this local law to promote the health, safety, and general welfare of the residents of the Town of Newfield, by the proper regulation and licensing of mobile home parks, to provide for a clean, safe, healthy and wholesome environment and living conditions within mobile home parks for the residents thereof.

SECTION 2.0

2.1   Section 2.1(C) of Local Law No. 2 for the Year 1989 shall henceforth read:

(C) "MOBILE HOME PARK" shall mean any parcel of land or contiguous parcels of land under common ownership, containing four (4) or more mobile homes, whether or not any such mobile homes are owned by the occupants thereof.

SECTION 3.0

Saving Clause, and Effective Date

3.1   SAVING CLAUSE

If any section, paragraph, subdivision or provision of this local law shall be adjudged invalid or held unconstitutional, the same shall not affect the validity of this local law as a whole or any part or provision thereof other than the part so decided to be invalid or unconstitutional.

3.2   EFFECTIVE DATE

This law shall take effect upon its filing with the Secretary of State in accordance with Section 27 of the Municipal Home Rule Law.

APPENDIX 6

Manufactured Home Community Rules and Rental Agreement

Appendix 6 is similar to a document of a respected manufactured home community. It combines the lease and the rules and regulations into one document.

MARCH 1, 1989

RESIDENTIAL TENANT REGISTRATION, RULES AND RENTAL AGREEMENT
PLEASANT HILLS MANUFACTURED HOME COMMUNITY

Your Address Will Be:

_____ Ithaca, NY 14850

DATE _____   HOME TELEPHONE _____

OCCUPANT NAMES   BIRTH DATE   EMPLOYER AND WORK TELEPHONE

_____   _____   _____

_____   _____   _____

_____   _____   _____

FORMER ADDRESS _____

MAKE-YEAR-LICENSE OF CAR(S) _____

MAKE-YEAR-SIZE OF HOME _____ CONDITION _____

NAME AND ADDRESS OF FINANCIAL INSTITUTION WITH LOAN ON YOUR
HOME _____

_____

IN CASE OF EMERGENCY NOTIFY _____ ADDRESS & PHONE _____

_____

TYPE OF APPLIANCE: RANGE: Gas   Electric   WATER HEATER: Gas   Electric
DRYER: Gas   Electric   FURNACE: Gas   Oil   Electric
Wood

FURNACE MODEL _____ AND MAKE _____

_____ We will provide a key for our home to be kept in the office.
_____ No key is available. We will stay home or make arrangements with a neighbor when it
is necessary for access for service or repairs to our home.

* I have had the lease options explained satisfactorily to me and:

_____ I DO NOT WANT A 12 MONTH LEASE   _____ I DO WANT A LEASE

I have read the Rules & Regulations and agree to comply with them, so that I may help my
Residential Manufactured Home Community become the most desirable in the area. It is
understood that the Rules & Regulations, as implemented by the Manufactured Home Community Owner, are incorporated in any lease +/or rental agreement. I am aware that Section
233 of the Real Property Law affects the Manufactured Home Owner and tenant relationship.

*applies to NEW manufactured home owner
tenants coming into the community.   _____

                                       (tenant's signature)

_____   _____
(Management Approval)   *(tenant's signature)*

PLEASANT HILLS RULES AND REGULATIONS   MARCH 1, 1989

I   MANAGEMENT RESPONSIBILITIES

Our Rules and Regulations are written for your general welfare, safety and enjoyment for pleasant and comfortable living. We plan to administer our rules fairly to all residents to assure that a few thoughtless people do not cause unnecessary inconvenience to you or to us. Rules are set to insure your comfort. To attain this end, we insist on compliance with our Rules and Regulations.

A. It is our intent to comply with the manufactured home owners Bill of Rights. We intend to cover the law properly within the following Rules and Regulations and to enforce the rules AND laws as outlined in Section 233 of the New York State Real Property Law.

B. LOT RENT INCLUDES:
1. County, Town, School and special district real estate taxes.
2. Adequate water supply from Bolton Point.
3. Municipal sewage disposal.
4. Hard top or dust controlled roads.
5. Snow removal on roads.
6. Common sense management.
7. Patios and walks.
8. Street lights as possible.
9. Operation of the communities in accordance with applicable Health Department Rules and Regulations, applicable New York State Rules and Regulations, applicable Town of Dryden Rules and Regulations and with the Manufactured Home Owner's Bill of Rights outlined in the NYS Real Property Law.
10. Management, interest in comfortable living for you, while maintaining a profitable operation.

II   SERVICES AND CHARGES

A. As a tenant in our community, service personnel may be available at the prevailing labor rates. In the white pages of the phone book, we have our main number 555-4444 listed. This number is to be used from 6:00 a.m. UNTIL 5:00 p.m. MONDAY THROUGH FRIDAY. THIS NUMBER IS TO BE USED ON SATURDAY FROM 9:00 a.m. UNTIL 12:00 NOON. **THE AFTER HOUR EMERGENCY NUMBER IS 246-4321** and is our Answering Service WHICH WE UTILIZE FOR <u>HEATING EMERGENCIES ONLY</u> from October 15th to May 15th and year round to report ANY EMERGENCY situations involving COMMUNITY OWNED electric and water equipment or systems. Our contract with the Answering Service is to notify servicemen of certain emergencies listed herein, during the hours NOT listed above, SUNDAYS AND HOLIDAYS. It is NOT our policy, or part of our contract with the Answering Service, to regularly see if there are any messages or requests for service. Their ONLY involvement with us is to notify servicemen of the above mentioned emergency situations when our office is closed. DO NOT CALL THE ANSWERING SERVICE and leave messages for us to call back or request different types of service. We DO NOT use the Answering Service for this purpose and we DO NOT get these messages. If you want prompt service for non-heating requests, we ask that you call our office during regular business hours, or you may leave a non-emergency message on our office phone number answering machines. We listen to these messages when we open the office on regular business days. If you expect service the same day, it is IMPORTANT THAT YOU CALL AS EARLY IN THE DAY AS POSSIBLE AND AS SOON AS YOU KNOW A PROBLEM EXISTS. We can provide good service to you ONLY when you make your request to us in the above businesslike manner and as soon as you realize you have a problem.

B. All community repair work and regular repair service and requests MUST BE MADE AND CLEARED through the office. Our personnel in the field ARE NOT authorized to accept service or repair requests.

C. If a tenant chooses to call another service company, and the problem is determined to be in our equipment, we accept NO RESPONSIBILITY for charges to the tenant by the other service company. DO NOT ALLOW an outside service company to tamper with any community owned equipment.

D. A resident may be evicted within three (3) days after receiving a written notice of delinquent rent payments.

E. The manufactured home owner agrees, by virtue of this agreement, to have the community owner, on the tenant's behalf, sign a UCC1 giving the community owner a security interest in his home for an amount not to exceed the balance owed the community owner for rent, services or other items provided by the community, and affirms that the community owner may file the same as a lien against the manufactured home whenever the community owner determines it is in his best interest to do so. Such filing shall not be released until the manufactured home owner has satisfactorily paid ALL amounts due the community owner.

F. The responsibility for payment of all amounts due the landlord shall be the joint and several obligation of the homeowners and those adult's names on the manufactured home owner tenant registrations. In the case of our rental unit tenants, responsibility for payments of all amounts due the landlord shall be the joint and several obligation of those rental tenant's names on the tenant registration, and/or lease.

G. Court appearance costs and/or landlord's legal expenses to collect past due bills or to enforce rules will be borne by the tenant in cases where the tenant has not corrected a violation of, or problem with the rules, a lease, any pertinent agreement, or past due bills within 10 days of the date of the notice of the problem.

H. If the community owner determines that it is in his best interest, the community owner may require the manufactured home owner and/or manufactured home tenant to provide a qualified individual(s) to guarantee any bills that the manufactured home and/or the manufactured home tenant might incur with the community owner or community service operator.

I. See current schedule and policy (SECTION XI) for additional details on costs, billing procedures and policies.

III   ENTRANCE, TRANSFER AND EXIT PROCEDURES

A. It is the intention of the community owner to limit occupancy in each unit to one family, and to those people on the original or amended registration for and/or lease filled in by the tenant, approved by us and on file in our office. It will be a violation of the Rules to have any additional people living in a particular unit without having been given written permission from the community owner, following a request by the existing tenant or manufactured home owner. All persons living in any unit MUST meet community residency criteria and follow ALL usual community entrance procedures including CREDIT CHECKS AND COMMUNITY RULE ORIENTATION in our office.

B. Reference forms, leases, rental agreements, credit reference forms, registration forms and all preliminary documents MUST BE completed by all occupants and approved by the community owner prior to the occupancy of a unit or lot. Non-registered home occupants will be dealt with as trespassers on our private property. The community owner may require updated tenant registration information from time to time. Existing tenants not completing and forwarding to us a signed, completed and accurate registration information form will be in violation of Community Rules. Community provided services may not be provided, or continued if ALL documentation necessary for occupancy has not been satisfactorily completed and approved by the community owner. No new tenant will be approved for occupancy of a particular lot if there are any unsatisfied rule violations relating to that home or that lot.

C. The community owner reserves the right to refuse new or continuing tenancy to anyone whose home is not considered acceptable in appearance or condition. Upon the proposed sale of a manufactured home the community management will do an inspection of the home, the accessories and the lot to determine if they meet community standards. If they pass inspection, the community owner will consider continuing tenancy for the buyer. If the home, lot and accessories do not meet the standards a list of what has to be done will be given to the present owner and the prospective owner. No such inspection will be done until we have received and approved a credit application from the prospective buyer. Inspections will be based on things being in good condition, on the provision that proper maintenance appears to have been done, and that community rules and standards are complied with.

D. "FOR SALE" OR "FOR RENT" signs may be placed in the windows or attached to the exterior of the mobile home, with no more than two 15" × 20" signs allowed on one home at a time. NO "FOR SALE" OR "FOR RENT" signs will be allowed at any other locations on the mobile home lot. Written permission is needed to install any other sign or type of sign anywhere in the community.

E. Mobile homes may not be rented, loaned or used by anyone for any other purpose than that granted in the original application for space rent, except with the written permission of the community owner. Residents ARE NOT permitted to sell or rent a home, with promise of occupancy, unless the new party is approved by the community management and fulfills ALL new tenant admission requirements.

F. The manufactured home owner MUST give the community owner twenty (20) days written notice of his intention to sell, rent, or remove his home from the premises.

G. The owner of any manufactured home in our community will be responsible for paying all bills due us on that home BEFORE occupancy can be approved and allowed for a new tenant, or BEFORE the home can be removed from the community. ALL RENTS, PARTS AND SERVICE BILLS AND MISCELLANEOUS CHARGES ARE THE RESPONSIBILITY OF THE MANUFACTURED HOME OWNER AND NOT his tenant in the case of a rented unit not owned by us.

H. Anything left in the community by the tenant shall be deemed abandoned and become community property as soon as the manufactured home leaves the community. ALL SHRUBS AND TREES BECOME THE PROPERTY OF THE COMMUNITY OWNER unless other arrangements are made that are satisfactory to the management.

I. Anyone wishing to rent their home must: 1) Secure approval of the community owner BEFORE offering their home for rent. 2) Secure approval of the community owner of each tenant. At least two business days are needed to process the written reference form and tenant registration forms required from each prospective tenant. 3) Be responsible for paying rent, miscellaneous charges and all parts and service work performed, as they will be CHARGED TO THE HOME OWNER AND NOT THE TENANT IN RESIDENCE. 4) AGREE TO OFFER HIS UNIT FOR 30 DAY LEASE PERIODS ONLY.

J. When an existing tenant leaves, we may shut off the water. NOT BEFORE WE APPROVE THE NEW TENANT, AND ALL BILLS ARE CURRENT, WILL THIS SERVICE BE RESUMED.

K. The community owner reserves the right to approve or disapprove all applications for tenancy based on character and/or credit references and/or other reasonable criteria that may from time to time be established by the community owner.

L. To vacate your lot, all residents MUST notify the community owner, in writing, twenty (20) days in advance of moving out. All outstanding accounts must be paid in full and a forwarding address given. Removal of a home will be done ONLY between the hours of 8:00 a.m. and 5:00 p.m. exclusive of Saturdays and Sundays and legal holidays. All removals must be supervised by our staff. A final day and hour of move out time must be made to our office at least 48 hours prior to departure, so that a final billing can be scheduled and all bills rendered and paid in full that day by the manufactured home owner prior to (1) moving the home off the space or (2) a new owner occupying the same manufactured home. FINAL BILLS ARE NOT MAILED SINCE ALL FINAL BILLS MUST BE PAID THE DAY RENDERED AT OUR OFFICE.

M. No approval will be granted for a new tenant to move in, or for a home to move out, until all bills due us are paid in full.

N. No manufactured home owner should infer to his buyer that the manufactured home may stay in the community without the approval of the community owner and the manufactured home owner must notify the prospective buyers that they will not be allowed to remove the home from the community or occupy the space if any bills due us are outstanding, or if entrance procedures are not followed.

O. Pleasant Hills Manufactured Homes will be pleased to act as sales agent for any residents wishing to sell their home. Check with the office for details.

P. State Health Codes apply to the number of occupants allowed in each dwelling unit.

Q. The landlord has not and will not charge manufactured home owners for RR ties used to support their manufactured homes. They are provided by the community owner, are his property and must remain in the community when a tenant vacates his lot.

R. FLOOR SPACE FOR SLEEPING ROOMS – In every dwelling unit of two or more rooms every room occupied for sleeping purposes shall contain at least seventy (70) square feet of floor space for the first occupant and at least fifty (50) square feet of floor space for each additional occupant thereof.

S. No additional three bedroom homes or third bedroom add-a-rooms will be allowed to be installed at PLEASANT HILLS after January 1, 1988. The three (3) bedroom homes that are presently here will be allowed to stay so long as they are occupied by the present owners. At the time of the re-sale of any three (3) bedroom in Pleasant Hills, it must either be converted to a two (2) bedroom (i.e., make one bedroom a den with no doors to enclose this room as a sleeping space) or the home must be removed from the community. Guidelines for converting three (3) bedrooms to two (2) bedrooms are available from the community management and will, in most cases, be much simpler than you might expect.

IV   HOME AND LOT MAINTENANCE

A. The tenant will be responsible for skirting his home within sixty (60) days after arrival in the community, except for those people arriving November 1st through March 31st. Their sixty (60) day period will begin March 31st. The tenant must provide access to all utility connections under the home in their skirting. The community owner cannot service utilities on units without access to the connections. Once skirting is installed, it must be properly maintained. All new skirting installation and replacements must be commercially manufactured home skirting made with metal or vinyl material. Any insulation or vapor or wind barrier must be installed behind the skirting panels.

B. Each manufactured home lot must be kept neat and clean. No regular and substantial storage of bottles, boxes, equipment or anything else will be tolerated on an unenclosed patio or around the home. All storage must be under the enclosed home or in an approved storage building, properly secured to the ground. To use your patio for storage, privacy paneling, meeting the approval of the community owner, must be installed. An approved storage building can solve your storage problems. All storage buildings must be kept in good condition, which means free from significant dents, properly painted and be satisfactorily maintained. All storage buildings will be manufactured or kit type, or commercially or professionally designed and assembled. Failure to agree to correct a shed problem within ten (10) days of notification may result in loss of permission to have a building. All buildings, sizes, and locations must be approved by the management in writing, prior to installation. After the effective date of these rules, no approvals will be given for more than one storage building per lot.

C. Awnings, carports, patio or screen rooms and expando rooms must be approved by the community owner. All carports and awnings and expando rooms must be commercially manufactured and approved in writing by the community owner. No homemade awnings or expando rooms or carports may be constructed on your lot. All awnings, carports, or expando rooms must be commercially manufactured and written approval from the management is necessary prior to installation. Screen rooms or patio enclosures that are professionally designed and assembled are acceptable following written approval from the management. The management requires plans or literature for the above be submitted before approval will be granted in writing. Screen rooms may be winterized on the INSIDE ONLY. NO PLASTIC OR OTHER MATERIAL may be applied to the OUTSIDE. ALL EXPANDO ROOMS MUST be capable of folding back into the manufactured home for transport OR be manufactured to be easily transported on their own chases or undercarriages. Carports, where possible, will be attached to the manufactured home. When not possible, carports must be commercially manufactured or free-standing style acceptable to the community owner. The above installations must be properly maintained.

D. Fences and landscaping around each home tend to make the community appearance more attractive. All fence installations, locations and material MUST BE approved in writing, by the community owner. A five (5) foot setback, from the road or parking area, is required for all permanent fences, shrubs or flower gardens. NO WIRE FENCES, EXCEPT ANCHOR TYPE CHAIN LINK properly installed and maintained will be allowed. Written approval from the community management is necessary PRIOR TO purchase and installation.

E. DO NOT DIG OR DRIVE STAKES into the ground without calling the office first. You might hit a buried electric, telephone or television cable or other utility line. The manufactured home owner will be responsible for damage caused to underground utilities by his or his tenant's actions.

F. ONLY umbrella style and retractable style clothes lines will be allowed for those tenants who wish to have their own outside lines on their lot. A tenant may install an approved clothes line ONLY after first receiving permission from the community owner as to its location and type. Clothes lines that do not have clothes on them must be folded up. Those who refuse this request may lose permission to have lines on their lot. No sockets, sleeves, pipes or pole may be driven into the ground for these clothes lines. All must be installed by hand digging the hole and then pouring a small bag of sacrete around the holder. No other type of laundry lines or poles may be installed on community lot.

G. Entrance steps and decks must be commercially manufactured, unless otherwise approved by the community owner, in writing. Steps must also be properly maintained. All occupants are responsible for cleaning and maintaining their own steps, decks and walkways and individual parking areas in a manner acceptable to the community owner and so that no hazardous condition is allowed to exist.

H. The tenant is responsible for maintaining, mowing and trimming his lot in such a way that it is neat and clean. Sites will be seeded ONLY ONCE by the community owner. Tenants are responsible for upkeep and maintenance including re-seeding and fertilizing. All tenants hereby recognize that they are, by virtue of continuing tenancy, authorizing the community owner to arrange for lawn clean-up and/or maintenance and/or lawn mowing whenever the community owner determines the lot and/or lawn needs attention. No notices will be given that this work will be done and the tenant will be billed at normal standard labor rates (see Section XI, C) and not at the voluntary lawn rate (see Section XI, B). Determination as to the need for the above will be at the sole discretion of the community management. We suggest if you cannot properly maintain your lot that you make arrangements with someone or that you request our voluntary lawn service.

I. Each lot may be improved by the tenant, or occupant (lawns, flowers, shrubs and trees, etc.). Annuals and perennials may be planted at the discretion of the resident. Hardy plants, shrubs and trees may be planted with written permission of the community owner following a written request and layout plan from the tenant. NO plantings are to be removed when the tenant vacates the lot without written permission from the community owner. DO NOT DIG MORE THAN SIX INCHES DEEP OR DRIVE STAKES AND/OR POSTS WITHOUT PERMISSION OF THE MANAGEMENT so that we may determine if there are underground utilities that may be jeopardized. Any damages caused by failure to do this will be charged to the manufactured home owner.

J. At any location in the community where the lawn extends to the off-door (rear) side of your neighbor's home, it will be the manufactured home owner's responsibility to maintain and mow that lawn to within three feet of the neighbor's home, carport, sidewalk, storage building and/or utility posts. It will be the home owner's responsibility to maintain that same three (3) foot area on the off-door (rear) side of his home. Other arrangements between adjacent neighbors are satisfactory, if those arrangements are agreeable to both neighbors involved, and the proper maintenance job is done. Mowing will be done to the front and rear of all lots to the first natural dividing area. If you are not sure of the proper area of your maintenance responsibility and if you and your neighbor cannot agree on what each of you will mow, call the office and we will show you. It is the management's intent to mow only a few common areas. Lawns are to be mowed only between 9:00 a.m. and 9:00 p.m. The management reserves the right, at any time, to determine what area should be mowed by any tenant. Anyone failing to comply with our reasonable request will be charged the applicable mowing rates, as described in Letter H above, for each mowing. The management does not provide any lawn equipment but does offer lawn service (see Section XI, B).

K. All occupants of our rental homes will be responsible for mowing their own lots unless the tenant has made arrangements with the management to have the management mow it on a regular basis and that cost will be reflected in the rent paid by such a tenant.

L. The community owner assumes the responsibility for plowing the community roads. Common shared parking areas will be plowed during heavy snowfalls only, and only if all cars are removed without notice from these areas as soon as the main road has been plowed, and before the plow is put away and only during regular business hours. If all cars are not removed from the parking area, these common parking areas will not be plowed. It is not practical to plow with cars in these areas. This parking area rule applies only to the areas where there are several tenant cars parking in one area. It does not apply to individual lot driveways or carports. Tenants will clean sidewalks, patios and individual parking areas. DO NOT use salt or calcium to melt ice on concrete sidewalks or patios. Any damage resulting from its use will be repaired by the community owner and charged to the tenant. The best solution for your slippery walks is to scrape the snow and ice clean. If you find you still need to use something else to solve an ice problem you may use sand or grits (available at Agway Feed Stores) or cracked corn (also available at Agway Feed Stores...the birds will also enjoy eating this). Tenants are responsible for maintaining their individual parking areas and their manufactured home lots in such a way so as not to create a hazard to themselves or others.

M. DO NOT leave faucets to garden hoses turned on when the hose is not in use.

N. Garbage pick-up will be one (1) day each week: Monday. Garbage must be out by the roadside by 7:00 a.m. All cans must be removed from the roadside the same day. Cans that are found out on any other day except Monday may be picked up and taken to the rear of the community shop and left there. The manufactured home owner or occupant will have to retrieve their cans from the shop themselves. The community owner will not accept any responsibility for cans that are picked up and placed behind the shop. On every manufactured home lot, there must be adequate containers to store refuse and so placed and maintained as to not create a nuisance. These containers will be purchased and maintained by the homeowner, not the community owner, except in the case of community owned rental units. We advise that you paint your name and/or lot number on your cans and tops. DO NOT place garbage at the roadside in open containers, cardboard boxes or paper bags. Garbage cans and properly secured plastic bags must be placed by the road on Monday ONLY. All refuse must be drained as free as possible of liquids. Garbage must be accumulated in durable non-absorbent, water tight containers having tight fitting lids with adequate handles to facilitate lifting. The interior of such containers will be kept clean by thorough washing and draining as needed. Heavy, water-proof garbage bags should be used inside the above containers. It is a violation of the Tompkins County Health Code to store garbage, refuse and rubbish improperly and create a health hazard with it. Trash pick-up service is available at the rates printed in the current rate schedule section of these rules. Any trash put out by the road on your lot will be assumed to be yours and you will be billed for its removal. Voluntary requests for trash removal must be called in to our office.

Recycling Rule – All storage and handling of items for recycling must be as follows: 1) All recyclables, in open top containers, must be stored inside your shed or inside your home, or may be kept outside in a properly maintained and tightly covered garbage pail supplied by the tenant. Garbage pails should be large enough to accommodate the two week recycling pick-up schedule. Recyclables stored in containers outside the home or shed must be in rodent-proof approved garbage containers. 2) Containers may be put at roadside on recycling days without the tight cover on the container. 3) It will be the tenant's responsibility to recover any items that blow, dump or spill out of your recycling container. 4) The recycle container must be out at curbside by 6:00 a.m. the day of recycling pick-up – watch our community notes as to a reminder of the pick-up days in the community you live in. 5) We recommend you paint your name on your recycling bins since you are responsible for them. The County may require a deposit or charge you for your bin if one has already been assigned to your lot, prior to your occupancy, and that bin cannot be found.

O. No additional TV antenna installations will be allowed in the community except for a satellite dish antenna. These may be approved on an individual basis. Tenants MUST MAKE A REQUEST IN WRITING with full details of what, where and how such a dish would be installed. None will be allowed without written permission from the community management. None will be approved if over ten feet in diameter or fifteen feet off the ground. None may be mounted on the manufactured home roof. The sighting on your lot must not have a negative impact on the area. Those wishing television service may contact the local cable vision operator for hook-up to their systems. Other antennas for radios may be allowed ONLY with written permission from the community owner after a written application from the tenant is submitted to and approved by the community owner.

P. Wood stoves will be allowed ONLY when the manufactured home owner uses listed and approved equipment suitable for installation in manufactured homes. Any new installations after the effective date of these rules must be done under permit from and inspection by the Town Building Code Officer and meet all UL and NYS equipment and installation requirements. All wood piles must be located in approved areas that are shielded either by trees, the manufactured home or fencing. These areas must be approved, in writing by the community owner, following a written request from the tenant including a map of the proposed wood pile location.

Q. All ashes from coal and wood stoves MUST BE stored on the tenant's lot in a tightly covered metal container, of not over thirty (30) gallon capacity until such time that they are removed from the community, by the tenant, to a proper dumping site. DO NOT SPREAD OR PILE THESE ASHES ON YOUR LOT, IN YOUR PARKING AREA OR ON OUR ROADS.

R. NO MORE THAN three (3) full cords of wood may be stored on the tenant's lot at one time.

S. To insure that your home can be moved, we require that you DO NOT sell or dispose of your axles and hitches.

T. Residents who wish to plant a small vegetable garden should request, in writing annually from the management, permission to have such a garden. The tenant's written request must include a map with proposed garden location and size of garden.

U. Manufactured home owners are responsible for obtaining necessary installation permits and/or certificates of occupancy for new home installations, modifications, wood stove installations and other situations where more than community owner approval is necessary. The community owner must first give his approval for any projects or installations prior to application by the manufactured home owner to the appropriate agency for construction or installation approval.

V. The exterior of manufactured homes and manufactured home lots must always be kept in a condition acceptable to the community owner. Improvements to the exterior of manufactured homes, expandos and manufactured home lots are encouraged by the community owner. All improvements must be approved in writing by the community owner before the improvements are started. All improvements must be maintained to the satisfaction of the community owner.

W. The manufactured home tenant and occupant will assume full responsibility for any pollutants such as heating fuel, motor oil, paint and any other contaminants that would create a hazardous or threatening situation to the environment. All problems caused by any type of spill or dumping of such items will be cleaned up by the tenant at his cost. The clean-up procedures are to be governed by the proper state or local agencies that handle this type of situation. The tenant will bear the full cost of any fines or clean-up needed as prescribed by the local, state and/or federal government. The cost of clean-up of any of the above mentioned pollutants and any fines that may be levied upon you by the proper government agency can be high.

X. All newspaper delivery tubes must be under the individual's rural style mailbox or attached properly to steps, decks or awning posts and not on individual posts by the road or in the tenant's yard.

Y. All homes must be numbered with commercially manufactured house/lot numbers that are easily visible from the street. Each letter and/or number must be at least 3" tall and less than 6" tall and at least 2" wide and less than 4" wide. They must be attached to the home or the awning or carport or an approved fence.

A-A. **Composting Bins** – Composting bins will only be allowed with written permission from the community owner or management. All composting bins must be commercially manufactured. No homemade bins will be allowed. Written request for approvals must be accompanied by a copy of the manufacturer's literature for the compost container you wish to install. Any existing unapproved or non-conforming composting operations must be removed by May 30, 1995. Locations of all composting containers must be approved and marked by the community management. All organic material placed in the composting bin must be shredded or chopped. In order to control odor and rodents, <u>NO</u> meat, dairy products and animal wastes will be allowed to be put into any composting bin. If there are any odor or rodent problems caused by your composting bin, you must purchase, at your expense, the proper chemicals, bait, traps, etc., needed to resolve the problem to the community owner's satisfaction or remove the bin from the community. Composting organic debris is a science – we strongly recommend that you obtain the newest available composting publication at the Agway Farm Store. If you are found to be in violation of these composting rules, you will be notified by the community management to remove the composting bin from your lot.

A-B. **Blacktop driveways** will be allowed to be installed on a tenant's lot at the tenant's expense. The manufactured home owner must get written permission from the community owner prior to installation of the blacktop. The community owner may wish to install, at his cost, underground sleeves or culverts or wires, or do some other infrastructure work prior to the area being blacktopped. This may minimize the need to dig up your blacktop to fix or install community owned utilities. In the event that your blacktop driveway was installed without written permission from the community owner, and repairs to any community owned utilities are necessary, any repairs to your blacktop portion of the driveway will be the sole responsibility of the manufactured home owner. If we had had the opportunity to approve the blacktop installation, we would have had the opportunity to review our infrastructure needs under that area prior to the manufactured home owner making the investment in the blacktop. The community owner accepts no responsibility for any damage done to the blacktop by utility companies.

V   CONDUCT OF TENANTS, CHILDREN AND GUESTS

A. All tenants will be held liable and responsible for behavior and conduct of themselves, of their children and guests, and will be held liable for damage done by any of the aforementioned people.

B. Children MUST be under the control and supervision of a qualified person at all times.

C. The use of profane, loud or boisterous talk WILL NOT be tolerated. Loud radios, televisions, stereos or other excessive noise is unnecessary. Disturbing noise is not permitted at any time.

D. Children are not to play in sections where there are no children their own age, unless they have received permission to do so from the tenant on the lot they wish to play on. Children are not to cut across other tenant's yards without permission of the tenant. It is the policy of the community owner that children are not to have the run of the community.

E. Children are not to play or congregate in the community roads.

F. NO CHILDREN will be allowed in, on or near any community owned buildings, oil tanks, pump houses or near repair, storage or construction areas.

G. Licensed motorcycles will be allowed, but are to be driven on the main roads with discretion and care and ONLY by properly licensed and equipped individuals.

H. The management is not responsible for damage, injury or loss by fire or theft to the residents' property, or the property of others associated with the residents.

I. The use of BB guns, firearms, or explosives on any community property is prohibited, unless authorized, in writing, by the community owner.

J. The pond in Pleasant Hills is an asset that can be enjoyed by many tenants. In order to share its values with all community tenants the following rules must be followed:

1. Fenced access points are provided by the community owner to allow tenants access to portions of the pond. These access areas are for the use of COMMUNITY TENANTS NOT having a lot that fronts on the pond. Tenants NOT LIVING ON THE POND are NOT to walk across tenant's lots that front on the pond.

2. NO SWIMMING IS ALLOWED.

3. NO MOTORIZED vessels are to be used on the pond. NO vessels may be left unattended in the water on the pond.

4. Fishing is allowed.

5. NO minor children are to be on or near the pond or pond dike, unless accompanied AND supervised by an adult responsible for these minor children.

6. Adult tenants wishing to ice skate must first take safety precautions and must make tests essential in determining the thickness and safety of the ice in the area they wish to use for this purpose.

7. All persons using the access areas must pick up after themselves and leave no mess of any kind.

8. NO campfires on the ground will be allowed at the access points.

9. Muskrats may be removed from the pond so long as the safety of tenants and pets is not jeopardized and State Laws are not violated.

VI   PETS

In a concentrated, residential area such as a manufactured home community, pets can be a nuisance. Pets are also a big responsibility. Carefully read over the rules regarding pets if you have one or are contemplating having one. You may love pets, but your neighbor may not.

A. NO DOGS OF ANY SIZE WILL BE ALLOWED EXCEPT A BONAFIDE SEEING EYE DOG FOR A CERTIFIED BLIND TENANT.

B. NO cats OR other pets will be allowed without written consent from the community owner.

C. A detailed application form, available from our office, describing the house cat or pet you want must be filled in. ONLY after receiving written permission back from the community owner may you obtain the pet. NO 'on the spot' or telephone decisions will be given. Granting permission for a pet, by the community owner, may take several days.

D. The number of cats per household is limited to two (2) ONLY.

E. ALL pet droppings on another tenant's lot must be picked up by the pet's owner immediately upon notification by anyone that his pet left droppings on that person's lot.

F. NO permission will be granted in our rental units for ANY DOG, CATS OR OTHER FUR BEARING OR FEATHER BEARING ANIMALS.

G. If you are found to be harboring an unapproved pet in your home, without permission, you will have to get rid of the pet and you will lose the right of future permission for six (6) months. You also could be evicted from the community as a rule violator.

H. If we determine that you are violating these pet rules or if your pet disturbs your neighbors, and a written legitimate complaint is filed with, and confirmed by, the community owner, you will have to dispose of your pet. Non-neutered cats causing problems and annoying other tenants will have to be removed from the community or neutered.

I. The only home that an approved pet has permission to live in is the home it was originally approved for. If a pet is going to move to another unit (lot) it must be approved for that lot prior to moving; otherwise, it will be considered an unapproved pet. Pet approvals are not transferable to other lots.

J. No cat pet applications will be approved unless a copy of the most current rabies vaccination certificate and a copy of any required licenses is submitted with the pet application. It is the law, in Tompkins County, that ALL CATS receive rabies vaccinations after they are three months old. It is the pet owner's responsibility to keep licenses and rabies vaccinations up to date as required by law. Pet approvals may be withdrawn if management finds pet owners are out of compliance with this or other pet rules.

K. Miniature pot-bellied pigs will be allowed only as a pet. NO breeding will be allowed and males must be neutered. Only one per household will be allowed. They must be kept inside your home at all times unless walked on a short leash. When walked off the owner's lot, the owner is responsible for picking up any dropping left on anyone else's lot or in regularly used community public areas. They will not be allowed to be tied out unattended at any time. All miniature pigs must be registered with the National Pot-Bellied Pig Registry Service and be a miniature breed not to exceed twenty (20) pounds adult weight. Special recommended diets, for the miniature pig, must be utilized to keep their weight down. If a miniature pig exceeds the weight, you can lose the right to have the pig in the community as a pet. All pigs purchased must be bred by a reputable breeder that has established his breed and the size of the pigs he has sold in the past; and, the breeder must be a member of the Registry Service. The community management requests a letter from the breeder stating the pig will not exceed the twenty (20) pound maximum weight limit, prior to approval. A pet application must be approved before the pig is allowed to come into the community. These pets will not be allowed in a household with any other non-confined pets. If the miniature pig creates a nuisance in your neighborhood you will have to remove the pig from the community.

VII   UTILITIES

A. Manufactured homes will be set up to perform to applicable New York State Manufactured Home Installation Codes, Health Department Codes, local codes, manufacturer's standards or any other standards applicable to the home being installed.

B. Manufactured home owners will bear full responsibility for installation and maintenance of plumbing, tie downs and utility systems (water, gas, oil, sewer and electric) from his home to the community receptacles or connections. A tenant should periodically make visual checks of these connections and report anything that looks 'out of order' to the community office. Tenants are responsible for electric and gas that goes through their meters. Since lines beyond the community provided electric connection, gas meter, the 4" sewer outlet and water riser pipe are owned by the tenant, it is the tenant's responsibility to periodically check and maintain them and to be responsible for any fuel or electric lost through their lines if there is a leak. The community owner's responsibility is only up to and including our water and sewer risers only and to our manufactured home electric service panel or junction box by your home.

C. DO NOT tamper with COMMUNITY OWNED utility installations, such as electrical boxes, community wiring, water curb stops etc. (EXCEPT to reset circuit breakers, replace fuses or turn off gas valves in an emergency). Should you have any trouble with any of these, please notify the office. Circuit breaker resetting and fuse replacement are your responsibility. You will be charged, at our regular service rate, if we are asked to reset breakers or replace fuses on your side of the electric meter.

D. Sewers are to be used for sewage and wastewater ONLY. No Kleenex, cigarette butts, paper towels, sanitary napkins, tampax, contraceptives or any other non-disposable substances including grease, garbage, coffee grounds, food particles or disposable diapers are to be disposed of through the sewer or drain lines. The lines cannot handle such items and will only cause sewers to back up. Costs of removing such items will be billed to the home owner.

E. The community owner reserves the right to periodically check your plumbing fixtures and outlets for wastefulness. A tenant's water supply may be shut off if the tenant allows water to run to keep pipes from freezing, has leaky faucets, a toilet tank valve that does not shut off, has any other loose connections or valves from which water is leaking or allows other unnecessary wasteful water practices. Services will be restored only after the community owner is satisfied that the leak(s) has (have) been repaired satisfactorily. These leaks will also cause your above ground sewer lines to freeze. Tenants will be charged regular service rates if we are asked to thaw out above ground sewer lines.

F. A properly installed heat tape is neither a fire hazard nor a large expense to operate. The community management asks that all heat tapes be installed to proper standards by the manufactured home owner. If, at any time, we determine that a heat tape is not in proper working order, the management will discontinue the water service until the tenant either has us replace the heat tape properly or he does it himself. There is no excuse for allowing water to run to keep your pipes from freezing. Heat tapes that are properly installed will protect your pipes from the weather conditions we have in this area.

G. All manufactured homes shall be installed with a water shut off valve in the line going to the home, that is accessible to the tenant so that he may turn off the water at any time he wishes to do so. It will be the manufactured home owner's responsibility to own and maintain this valve. It is not the policy of the community owner to have water keys available for shutting off underground valves or to have someone in attendance on a regular basis to shut off water. It is your responsibility to have a valve installed so you can shut off your own water in case of emergency or, when you need to make repairs to your water system.

H. Lawn and/or garden watering will be allowed only with automatic sprinklers or controlled hose nozzles.

I. Oil and Gas Installations – To avoid unsightly individual oil or LP gas tanks on individual lots, the community has made available natural gas by means of a metered underground delivery system. On any lot where the manufactured home community owner has provided access to the community's central natural gas system and whereupon the manufactured home owner chooses not to purchase gas for heating from this system, the manufactured home owner, at his own expense and with written permission as to the location and other criteria some of which is outlined below, may install his own oil storage equipment to service the manufactured home owner's home ONLY.

1. As of October 31, 1988, there will be no new LP tanks allowed to be installed in our manufactured home communities. As of July 1, 1989, all LP tanks, except twenty (20) lb. cylinders used for outdoor cooking, will have to be removed from the communities.

2. Direct burial oil tanks having a minimum capacity of 285 gallons, made of twelve (12) gauge or heavier steel, bearing underwriters label 'STP-3' and having a ten (10) year warrantee, may be buried on the manufactured home lot at a location approved by the community owner. The tank hole must be hand dug and back-filled with fine sand or bank run gravel with no stones larger than one inch in diameter. No power equipment will be allowed to be used to make the hole or the ditch for the lines for any oil line installation. Lines from the tanks to the home must be in a conduit and/or surrounded by sand. A properly installed, painted and supported 275 gallon above ground fuel tank may be installed in a tenant's storage shed or be enclosed

totally with any approved stockade fence at a location acceptable to the community owner. All oil tank installations must also be in conformance with appropriate local, state and national codes. No installations will be allowed without written approval from the management following a written application for such installations to the community owner by the manufactured home owner.

3. The management will cooperate so as to allow such installations so long as they do not cause either higher costs for the community owner or detract from the looks of the community. It is our intention not to have any above ground LP fuel tanks in the community.

4. Tenants using a kerosene portable heater must store their fuel in federally approved containers of not more than six (6) gallon capacity in a safe place so as not to create a hazard or detract from the appearance of the community.

5. The tenant and manufactured home owner agrees to assume full responsibility for any damage or problems caused by the installation or use of their natural gas or oil equipment.

6. If you think a gas appliance or a gas supply line has developed a leak, turn off all flames and open all windows and turn off any gas valves available to you and call for help. The NYSEG number for natural gas problems is 1-800-521-5572 for community tenants and is 347-4131 for Pleasant Hills tenants.

J. All gas and electrical connects and disconnects including electrical and gas service are to be arranged through the community office. Contact the office before purchasing washers, dryers, air conditioners, etc., to see if your site is adequately equipped for the installation of such appliances.

K. Those of you who leave the area for a warmer climate in the winter, or extended vacations, should let the office know if you want your utilities turned off and when they should be turned on again (see L below).

L. Proper winterization of your home is essential to keep pipes from freezing when there is no heat on in your home. Winterizing service is available at normal labor and material rates.

M. Manufactured home owners/tenants should familiarize themselves with the locations and operation of gas, oil, electrical and water shutoffs, so in case of any emergency they can turn off the supply until the problem is resolved. We will be glad to show you where they are on your lot.

VIII   VEHICLES, ROADS AND PARKING

A. The speed limit is 10 mph or as posted. Excessive speed will not be tolerated. Speed bumps do not damage slow moving vehicles. Everyone MUST comply with posted traffic control signs.

B. Cars, boats, campers, motorcycles, snowmobiles, utility trailers, etc., which are either unlicensed or not being used must be stored in an area approved, in writing, by the community owner, if such an area is available, or removed from the community premises.

C. NO major overhauling or other major repairs will be permitted on or around any manufactured home lot, in parking areas or roadways. Vehicles may not be left unattended while on jacks or blocks.

D. All parking will be in provided parking areas or in individual driveways. Guests may use common parking areas, in other sections, if there is no room in their host's parking area. Vehicles parked along the roads, on lawns or not in designated parking areas may be towed away without notice and at the discretion of the management, at the vehicle owner's expense.

E. No mini-bikes, go-karts, snowmobiles, three or four wheelers, motor scooters or all terrain vehicles will be allowed to be operated on our property unless the operator has an operator's license, satisfactory safety certificates or there is, on the site, adult supervision whenever a nonlicensed or noncertified driver is riding one of the above mentioned vehicles. Speed limits will be observed at all times. The above mentioned vehicles will be operated ONLY on community roads, or in the play areas, if any, at reasonable hours.

F. If you are leaving a licensed vehicle in the community during vacation time, please leave keys with a neighbor, or at the office so the car can be moved if necessary.

G. All vehicles that are allowed to be left in the community must be roadworthy, have current plates and valid registration and a current inspection sticker on them. Vehicles that do not meet the above criteria or are in the process of having major repairs performed on them or that violate other community rules will be towed at the vehicle owner's expense 10 days after we place a violation notice on the vehicle.

H. No unlicensed and/or uninsured motorcycles or motor vehicles are allowed to be operated on community property.

I. No vehicles may be left unattended on community roads at any time. This could cause a disastrous situation of ingress or egress, especially for emergency vehicles. Vehicles parked in roadways may be towed immediately and the ten (10) day notice provision for vehicles improperly parked will not apply to vehicles parked on roadways.

J. All vehicles of all occupants of all homes must be registered with the community office. It will be assumed by the community owner that any vehicle not registered with us, does not belong in the community and when problems arise, said vehicles will be handled as if they were owned by strangers and abandoned on our private property.

IX   COMPLAINTS

A. A manufactured home community is a group of manufactured homes located in close proximity to each other. At times minor problems will arise between neighbors. These problems should first be talked over and an agreement arrived at by these same neighbors. If you are not able to jointly solve the problem, after a reasonable attempt has been made by you, a WRITTEN COMPLAINT to the community owner would be in order. WRITTEN COMPLAINTS are handled by us with written responses relating to the rule violation.

B. If you have other complaints worthy of a written rule violation notice from us, present a complaint in writing to the community office.

C. COMPLAINTS WILL NOT be listened to, accepted by, or acted upon by any service employee while they are working in the community. COMPLAINTS MUST BE DIRECTED TO THE OFFICE.

X   GENERAL COMMUNITY POLICIES

A. Office hours will vary. Those in effect at any time, are posted on the outside of the office. Bookkeeping personnel to answer complicated questions regarding your accounts are available Monday through Thursday only. However, bills may be paid during Friday and Saturday hours. DO NOT EXPECT DETAILED EXPLANATIONS OF YOUR ACCOUNTS ON FRIDAYS OR SATURDAYS UNLESS A PRIOR APPOINTMENT HAS BEEN AGREED TO BY US.

B. We strongly urge that you consider carrying liability insurance and property damage insurance on your home and your personal belongings.

C. These Rules and Regulations are made with the intent to make this community a real asset to the Ithaca community, and a place in which you will be proud to live.

D. If you have any problems interpreting these rules, or if you want to do something that is not covered, it is necessary to consult the office before going ahead and doing it. This may save us both problems.

E. A resident will be given ten (10) days to correct a violation of community rules after we have delivered a written notice. If he fails to correct said violation, he will be notified to vacate within thirty (30) days.

F. Any expenses incurred by the community owner because of negligence or noncompliance with the community rules, by the tenant, or manufactured home owner, will be the responsibility of the manufactured home owner except in the cases of our rental units, the tenant will be responsible.

G. The community owner reserves the right to add, change or modify these Rules and Regulations from time to time. These Rules and Regulations may be changed after 30 days written notice by community owner to tenants. A tenant who does not agree with a new rule or regulation may choose to give the community owner notice that he intends to leave the community prior to implementation of said rule or regulation or fee schedule. By virtue of a tenant remaining in the community after effective date of any rule, regulation or fee schedule, the tenant agrees to abide by such rule.

H. For items that are not covered specifically in the rules, the management reserves the right to make reasonable, common sense policies that affect the situation and to implement them immediately.

I. These Rules and Regulations are necessary in order to have a neat, clean, orderly and enjoyable facility. An infraction of these Rules is adequate grounds for eviction, as defined in the New York Real Property Actions and Proceedings Law. The community owner does not want to evict anyone, but violations of these rules may make such action necessary.

J. A violation of some of these rules is not only a violation of these community rules, but could be a violation of the Tompkins County Health Code or other local or state codes or laws. Judicial action and prosecution is possible for noncompliance in these areas.

K. These Rules and Regulations are provided for the convenience and welfare of the residents. We realize that some may cause inconvenience at times, but, if you desire to live in a clean, well kept community, rules are necessary.

L. Where written approval is required throughout these rules, it is assumed by the management that no approvals will be given until after we receive a written request from the manufactured home owner or tenant.

M. Tenants and manufactured home owners agree that they will come to the community office, when requested to answer concerns or problems that the manufactured home community owner may have with them or their home or lot.

N. No commercial enterprises are allowed on your manufactured home lot or in your manufactured home without the written permission of the community owner.

O. Although our staff is in the community on a regular basis, it is not possible for us to see or hear about every violation of community rules; it is our intention, through these rules, to make our communities the best in the County. In order to reach this goal, your assistance is necessary in letting us know what rules are not being followed, and by whom. Everyone must accept some responsibility in order that our communities are the best in the County. May your stay here be an enjoyable one. Please feel free to offer suggestions for the improvement of this residential community.

XI   CURRENT RATE SCHEDULE AND POLICY

A. The lot rental schedule is as follows: All rent is DUE on the DUE DATE, and payable in advance. To qualify for the net rental, ALL PAYMENTS MUST BE RECEIVED IN OUR OFFICE OR POSTMARKED NO LATER THAN THE 10th day into the month (excluding Sundays or Holidays falling on the 10th) and must be paid in negotiable funds. Rent must be mailed or brought to our office. An after hours deposit box is available at our OFFICE for your convenience. DO NOT PUT CASH IN THE DEPOSIT BOX OR IN THE MAIL.

LOT RENTAL SCHEDULE EFFECTIVE APRIL 1, 1996	DUE DATE	GROSS RENT	NET IF PAID BY THE 10TH OF THE MONTH
10 or 12 wides	1st	$209.00/mo.	$194.00/mo.
14 wides	1st	$224.00/mo.	$209.00/mo.
Double wides	1st	$254.00/mo.	$239.00/mo.

SENIOR CITIZENS, VETERANS OR CLERGYMEN MAY QUALIFY FOR LOWER RENT. CALL OUR OFFICE FOR DETAILS. A rent reduction is possible when Senior Citizens, Veterans or Clergy exemptions are properly filed with and approved by the Tompkins County Department of Assessment. They will be honored and credits given for applicable tax periods covered by these exemptions. They MUST be done YEARLY and BEFORE MARCH 1st. Do not wait until the last minute to file to see if you qualify.

B. Voluntary Lawn Service – A rate of $20.00/hr. ($10.00 minimum charge – and time will be charged in ¼ hour increments after the first one half hour) will be charged for lawn and/or lot clean-up labor, when mowers, vehicles and/or tools are provided by the community owner. This is for *voluntary* lawn clean-up work only. Any involuntary lawn maintenance will be charged a minimum of one hour of the standard labor rates below. See Section XI, Rule C.

C. LABOR RATES

1. STANDARD LABOR RATES – for work requested so that it can be scheduled and accomplished between 8:00 a.m. and 4:00 p.m. Monday through Friday (except holidays when we are closed)

$40.00/hr. for first man for labor and travel
$30.00/hr. for each additional man on the same job
Minimum charge is one hour per man on job

2. OVERTIME LABOR RATES – for emergency work requested so that it must be done between 4:00 p.m. and 8:00 a.m.* or on a Saturday* or Sunday* or holidays when our office is closed.*

$60.00/hr. for first man for labor and travel
$45.00/hr. for each additional man on the same job
Minimum overtime charge is 1 and ¼ hours of travel and labor ($75)

*WE RESPOND DURING THESE TIMES TO NO HEAT CALLS, FROM OCTOBER 15TH TO MAY 15TH, OR BROKEN WATER OR ELECTRIC EQUIPMENT OWNED BY THE COMMUNITY

3. Time for standard labor rates is computed from the time the man leaves our office until he returns to our office from the job and includes parts acquisition time, except:

a.) No more than ½ hr. travel time will be charged for tenants in our communities when the Standard Labor Rate is in effect.

b.) Overtime labor rate will be charged from the time the man leaves his home until he returns home including any parts acquisition time.

D. GENERAL BILLING POLICY

1. A ten percent (10%) Senior Citizen Discount is available to persons over 60 years of age on parts and labor to a maximum combined discount of $20.00 per sale. THIS DISCOUNT MUST BE REQUESTED WHEN THE WORK IS REQUESTED. None will be honored after the work has been completed and bills rendered.

2. Any time anyone becomes delinquent on his accounts owed us, a security deposit may be required from that individual. Failure to pay that security deposit within two (2) weeks after requested will be a violation of community rules, and could mean a termination of services or products provided in the area of unpaid accounts or unpaid security deposits.

3. Past due charges of 2% per month ($.50/mo. min.) will be added to Parts and Service and Miscellaneous charges that are 30 days or more past due.

4. All payments or credits to your account will be applied to the OLDEST BALANCE DUE US ON YOUR ACCOUNT. The following are due dates for various categories of bills: All rents are due on the 1st and all rents not in our office by the 10th day into the month (excluding Sundays or Holidays falling on the 10th of the month) or postmarked by the 10th will not qualify for the rental discount and the full gross rent amount will be due. Parts and/or Service and/or Miscellaneous charges are due on the dates completed and billed and not 30 days later. When applying money received, all money will be credited to the oldest balance due using the above rate schedule. This means that the tenants are not able to pay selectively on their account as we will be applying any payments to the oldest account first, based on the due date schedule above.

5. A $15.00 charge, plus tax, will be made for each returned check each time it is not honored, for any reason, and is returned to us unpaid by our bank.

6. Duplicate bills can be mailed, at no charge, to a second party at the request of the manufactured home owner.

7. If proper notices relating to vacating the premises are not given us by tenants and/or if new tenant entrance procedures are not complete by the 10th of any month, the next month's lot rent will be computer billed to the tenant in residence on the 20th of the month.

8. Bills will be rendered every month.

9. If the landlord determines that it is in his best interest, the landlord may require, at any time, the manufactured home owner/tenant to provide a qualified individual to guarantee any bills that the manufactured home owner might incur with the community owner or operator.

TRASH DISPOSAL FEES – EFFECTIVE JANUARY 1, 1991

Flat rates below include pick-up at your roadside and disposition by us at Landfill or appropriate disposal area. "T+M" indicates we charge for time, material, hauling and disposal costs for these items marked "T+M".

Metal Shed (Unassembled)	$40	Tires	$10/Tire
Metal Shed Assembled)	T+M	Building Materials	T+M
Wood Shed (Unassembled)	$80	Dinette Table and/or	
Wood Shed (Assembled)	T+M	Chairs	$15/Set
Washer	$30	Living Room Chair	$25
Dryer	$30	Living Room Couch	$40
Dishwasher	$30	Carpet	$35/Room
Range	$30	Bed (or part of)	$30
Furnace	$30	Brush generated on	
Refrigerator or Freezer	$60	your own lot	No Charge
Room Air-Conditioner	$30	Untagged Garbage	$20/Bag
Hot Water Heater	$30	Television	$20
Lawn Mower	$20		
Old Skirting	$50	WE DO NOT PICK UP	
Batteries	$10		
Bicycles	$10	PAINT, CHEMICALS, HAZARDOUS	
Car Parts		MATERIALS, OIL, GRASS CLIPPINGS,	
(minimum charge – one garbage		LEAVES	
can full)	$25/20 gal.		
	garbage can		

For items not listed we will charge the same as a similar item on above list.

RULE AMENDMENTS – PLEASANT HILLS RULES AND REGULATIONS

Effective Date	Section – Paragraph	Rule Contents
05/20/91	III – S.	The first, second and third sentences of this rule are hereby deleted and the following language is inserted: "No additional three bedroom homes or bedroom add-a-rooms or conversion of existing two bedroom homes to three bedroom homes will be allowed after May 20, 1991."
04/01/93	II – I.	Repeal rules relating to garbage credit.

A Copy of an Eviction Notice Used by a Manufactured Home Community

PAST DUE RENT & EVICTION POLICY

It has become necessary to drastically change our policy toward those few tenants who are past due on their rents. It is not fair to those tenants who pay their rent on time to have to pay higher costs for those few who are past due. We have implemented a new policy that will deal with past due rent differently and will minimize the costs that have to be passed on to tenants who pay their rent on time.

We also have determined that a space is better off empty than to have it occupied by a tenant who is not paying their rent since our costs go on if we continue to provide them with water, sewer, community services as well as pay the taxes on their home.

THE NEW POLICY IS AS FOLLOWS

(1) If a tenant gets 45 days past due on any rent balance due us we will expect the following
   A. The tenant must meet with us and set up a definite payment schedule that will allow them to get current with us in a reasonable length of

time. This repayment schedule must be acceptable to us and must be carried out as we and the tenant have agreed.

B. If the tenant does not meet with us to set up such a schedule; and/or does not carry through with an agreed on catch-up schedule; and/or does not keep us updated on any change in their ability to carry out their responsibility as agreed, we will have to begin the procedures outlined below.

C. Eviction starts with us giving the tenant a 3 day demand for payment in full of the account. If this is not paid in the 3 days we have a process server serve the uncooperative tenant with a notice to appear in the local Town Justice Court. We present the facts to the Court and the Judge makes a decision. Historically, local Court Judges have little patience with people who are attempting to get a free ride at the expense of others. As this procedure goes on, the past due tenant will be charged all of our legal, court and administrative costs since the community rules agreement they signed allows us to charge them for all our costs.

D. If the tenant has not moved his home from the community by the date set by the Court, the Sheriff will supervise the removal of the home to a storage facility. To retrieve the home the tenant must pay all the extra costs associated with the removal and storage of the home and all the other past due charges including all the back rent charges.

E. As soon as the home is removed from the community, we summarize all the money due us for back rent, court and legal charges, and removal expenses we incur, and turn this amount over to a national credit bureau who starts formal collection proceedings. If the ex-tenant does not respond to requests to pay, they sue the past due tenant and get a judgment that is recorded in the County Courthouse until it is paid. This credit and eviction record becomes public and will make it more difficult to get credit to buy a car or even rent someplace else.

Our goal is to have as many spaces occupied by rent paying tenants, but it is better for us to have the site empty if there is no rent coming in from it. Very few of you will ever get to sections B, C, D, or E above, especially if paragraph A above is followed to the letter.

APPENDIX 8

Standard Drawing Details

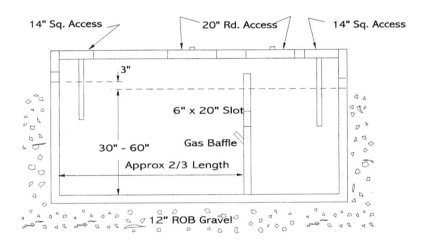

14" Sq. Access

20" Rd. Access

14" Sq. Access

3"

6" x 20" Slot

Gas Baffle

30" - 60"

Approx 2/3 Length

12" ROB Gravel

Figure 1
Precast concrete septic tank.

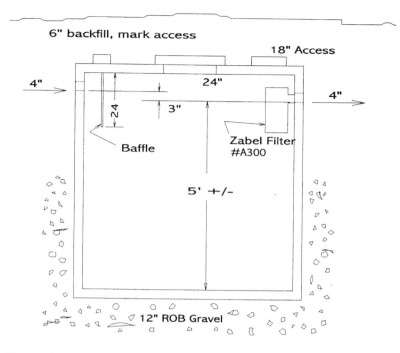

Figure 2
Tank with Zabel filter.

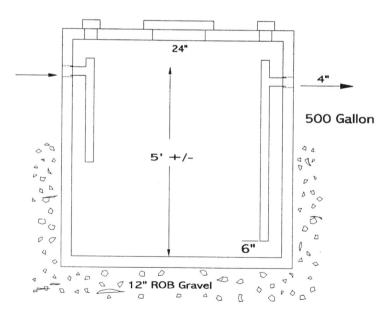

Figure 3
Grease tank sketch.

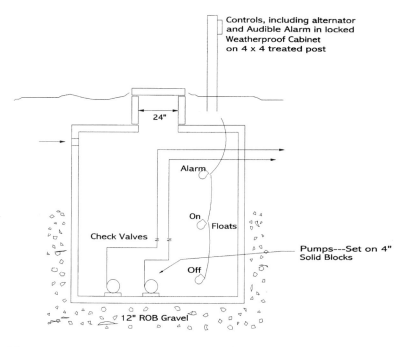

**Controls, including alternator
and Audible Alarm in locked
Weatherproof Cabinet
on 4 x 4 treated post**

24"

Alarm

On
Floats

Check Valves

**Pumps---Set on 4"
Solid Blocks**

Off

12" ROB Gravel

Figure 4
Pump tank sketch.

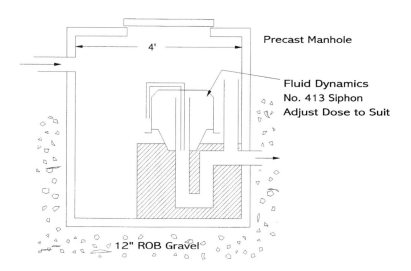

Precast Manhole

4'

**Fluid Dynamics
No. 413 Siphon
Adjust Dose to Suit**

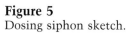
12" ROB Gravel

Figure 5
Dosing siphon sketch.

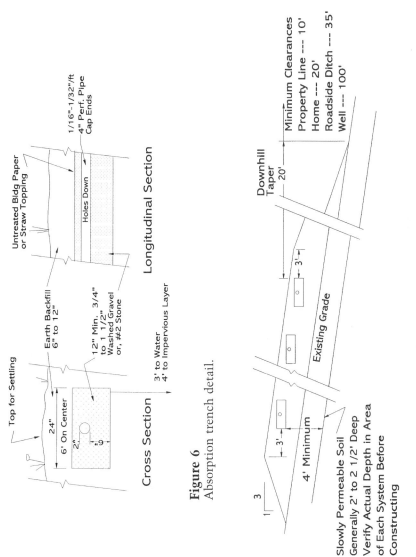

Figure 6
Absorption trench detail.

Figure 7
Elevation detail, raised bed systems.

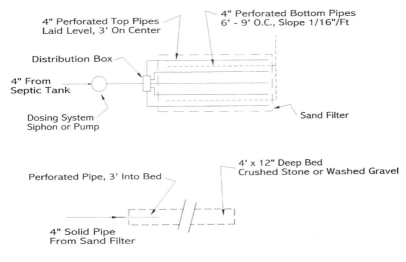

4" Perforated Top Pipes
Laid Level, 3' On Center

4" Perforated Bottom Pipes
6' - 9' O.C., Slope 1/16"/Ft

Distribution Box

4" From
Septic Tank

Dosing System
Siphon or Pump

Sand Filter

Perforated Pipe, 3' Into Bed

4' x 12" Deep Bed
Crushed Stone or Washed Gravel

4" Solid Pipe
From Sand Filter

Figure 8
Sand filter and absorption bed plans.

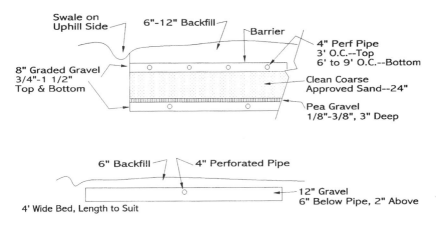

Swale on
Uphill Side

6"-12" Backfill

Barrier

4" Perf Pipe
3' O.C.--Top
6' to 9' O.C.--Bottom

8" Graded Gravel
3/4"-1 1/2"
Top & Bottom

Clean Coarse
Approved Sand--24"

Pea Gravel
1/8"-3/8", 3" Deep

6" Backfill

4" Perforated Pipe

12" Gravel
6" Below Pipe, 2" Above

4' Wide Bed, Length to Suit

Figure 9
Sand filter and absorption bed sections.

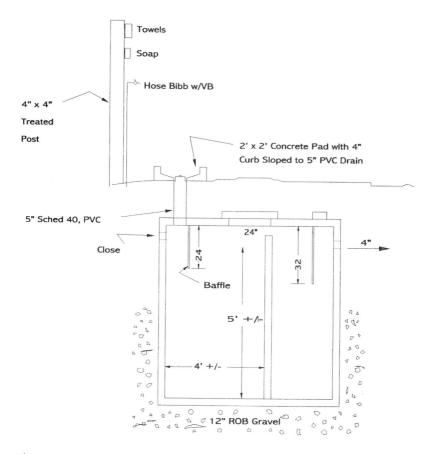

Figure 10
Dump station septic tank sketch.

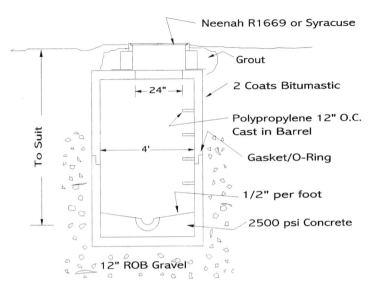

Figure 11
Typical manhole.

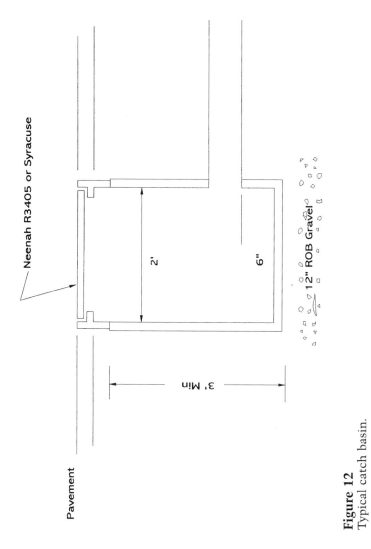

Figure 12
Typical catch basin.

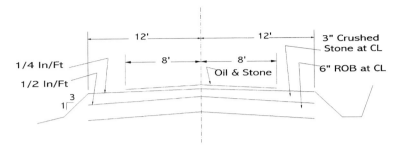

Figure 13
Road section.

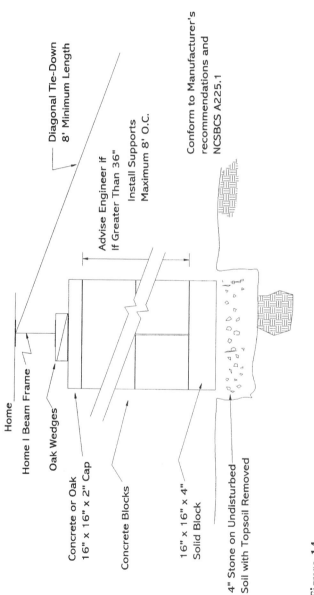

Home

Home I Beam Frame

Oak Wedges

Concrete or Oak
16" x 16" x 2" Cap

Concrete Blocks

16" x 16" x 4"
Solid Block

4" Stone on Undisturbed
Soil with Topsoil Removed

Diagonal Tie-Down
8' Minimum Length

Advise Engineer if
If Greater Than 36"

Install Supports
Maximum 8' O.C.

Conform to Manufacturer's
recommendations and
NCSBCS A225.1

Figure 14
Manufactured home support sketch.

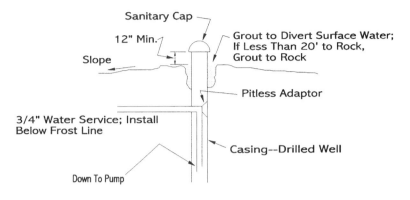

Sanitary Cap

12" Min.

Grout to Divert Surface Water; If Less Than 20' to Rock, Grout to Rock

Slope

Pitless Adaptor

3/4" Water Service; Install Below Frost Line

Casing--Drilled Well

Down To Pump

Figure 15
Well sketch.

Plan Notes

General

- Site data and topo are from survey by Robert Russler, Land Surveyor, information from Owner and field observations.
- Site data are from survey by Robert Russler, Land Surveyor.
- Site data are from tax map.
- Topo is from United States Geological Survey (USGS).
- Field measure and verify all data.
- All work shall conform to local ordinances, the Your County Department of Health, NFPA 501A, NCSBCS A225.1, NEC and manufacturers' recommendations. All material and equipment shall be suitable for intended service. Advise engineer of any conflicts. Advise health department of start of construction, and obtain its continuing approval as the project progresses.
- Prior to the start of construction of each of the septic fields, with the assistance of the Your County Health Department, inspect actual soil conditions in the exact area of each system. Secure sewage system construction permits for each lot from the Your County Health Department based on details provided on this plan; the Department will modify the size of each system based on actual soil conditions and actual number of bedrooms. The details shown on this drawing are to establish the general parameters, that adequate space exists for a septic system, a replacement system, a well, and the home.
- Obtain approval from the Your State Department of Transportation for road cut into NYS Route 96B.
- Advise engineer and request visits during critical periods of construction to satisfy health department's concern that the project is in conformance with the approved drawings.

- Provide site stand and anchors for all single wide homes per the latest edition of the manufacturers' instructions, or in accordance with NCS-BCS A225.1. Community owner to provide anchors. Home owners are responsible for installation and maintenance, including seasonal adjustments, of all tie downs.
- Owner shall not permit grease to enter the sewerage system. The septic tanks shall be cleaned and baffles inspected annually. Suggest annual inspection of the distribution boxes. Inspect the Zabel filter monthly until a frequency of cleaning can be established.
- Locations of homes are not restricted to "exactly as shown". Maintain specified separation distances from septic systems.
- Owner shall deliver specification sheets and equipment maintenance sheets to the engineer, including pump curves and similar data, for all mechanical operating equipment actually installed for preparation of the Operating Manual by the engineer. The Operating Manual will be required by the Health Department to obtain approval to occupy homes.
- Install 3 gpm flow restrictors on all showers. Check all water closets for leaks and presence of shower flow restrictors on a semiannual basis.
- A fill system in an area with only minimum permeable soil can be expected to be soggy at times. Conform to NEC and provide certificate of electrical inspection.

Basis of Design

General

Design flows in the subdivision are based on 150 gallons per day per bedroom. For the purpose of this general layout, four bedroom homes are assumed for each lot. Design flows are based on 1988 DEC regulations for manufactured homes. All homes in the project will be double wides, and are considered as three bedroom units. The 20% reduction for low flow fixtures is applied.

Water

10 three BR homes @ 450 gpd	= 4500 gpd
Less, 20% low flow allowance	= 900
Daily water flow design	= 3600 gpd

Sewer

Each home on separate system	450 gpd
20% flow reduction not applied	

Water Supply

Install well and provide well logs to Health Department.

Two 1000-gallon water storage tanks presently exist. Add one 1500-gallon water storage tank. 3600 gpd equals 2.5 average gpm. Design with 25 gpm peak.

Provide one new 250-gallon bladder tank with 75-gallon pumpdown capacity from 50 psi to 30 psi.

Operate well pumps on storage tank SJ Electro floats and Grainger 2E357 cycle timers, 5 minutes on, 10 minutes off.

Pump Head Required

Apartment building elevation, feet	1155	
Storage tank elevation		1100
Static head		55
Top floor of building		20
Min pressure of 25 psi =		60
Total pump head required	135	

Measure water level in well, start pump and pump to drain, measure water level in well after 30 minutes, 60 minutes, and every hour. After water level stabilizes, continue pumping for four hours and measure yield.

Sewage Disposal

At the start of construction of each septic system, inspect actual soil conditions in the exact area of each system. Design is based on percolation tests of 45 minutes. Advise engineer and County Health Department if actual conditions vary. Groundwater or impervious layer must be at least four feet deeper than bottom of absorption trenches.

450 gpd/0.5 gpsfpd = 900 square foot trench area required

900/2' wide trench = 450 lineal feet of trench required

Provide five 90' long, 2' wide trenches

Add notes if dosing is required.

Provide gravel/loam fill to obtain a separation of at least four feet from the bottom of the tile trenches to groundwater or impervious layer. Fill to perc at rate between 10 and 45 minutes per inch. Obtain Your County Health Department approval before placement. Compact fill in 8" lifts with light bulldozer only. Obtain Health Department approval after placement but before installation of trenches.

Provide two 18 feet by 35 feet sand filters and alternately dose. Filter sand shall be with an effective size of 0.25 to 0.6 mm and uniformity coefficient of ≤4. All sand must pass a ¼ in. mesh screen.

Dispersion trench to be based on actual expected flow of about 450 gpd from each sand filter. A loading of 1.2 gpdpsf results in two 3 feet by 65 feet long trenches needed for each filter. Trench shall be similar to tile field without perforated pipe graded at a slope of ¹⁄₁₆ to ¼ in. per foot. Trench must follow ground contours, without pockets, to allow free flow through the entire trench. Trenches shall be a minimum of 50 feet from property lines, ultimate discharge to remain on property.

Material

New septic tanks shall be Keystone or Kistner consisting of two 1000-gallon tanks as indicated. Provide the A300 Zabel filter on the outlet of the final tank. Preliminary tank to have 24" baffle. Septic tanks shall be concrete as manufactured by Keystone or Kistner and in accordance with 1988 DEC regulations. Round septic tanks with internal baffles are acceptable for this project.

Provide 500-gallon tank with Zabel filter.

Dosing pump well shall be 575-gallon concrete tank with two Myers S-25 pumps, 24 gpm at 12 feet head. Set for 240 gallons per dose.

Provide Myers S-4 effluent pumps, with alternator, in 500-gallon effluent well for dosing of septic fields. Pumps to have capacity of 30 gpm at 20' head. Set dose at 300 gallons.

Water service shall be NSF approved PVC piping. Install Mueller or Hayward schedule 80 curb stops at each branch connection with 3/4" Type K copper service to each home.

Storage Tank — approximately one day's water consumption. One 8' round tank, 10' long, 3800-gallon capacity. Line tanks with NSF approved lining. Tanks shall be Highland with approved water lining, Koppers 654 and Koppers HB finish epoxy lining. Provide 24" manhole and connections to suit and cathodic protection. Tanks to be STI-P3.

All valves for the project shall be full port ball valves.

Water meter shall be a Badger 1½ in. meter with test port within a Ford Customsetter 2 in. assembly. Provide ball valves.

Blowoff valve shall be a ¾ in. Eclipse No. 88 Sampling Station enclosed in a lockable aluminum-cast housing.

Sanitary sewers shall be PVC, SDR 35, sized per the drawing.

Pumps — two, 15 gpm @ 150' each. Operate pumps on pressure switches. Pumps are sized for 50% of maximum peak flow with pressure tank. Head is selected at 45 psi minimum discharge pressure. Highest home in complex is approximately 30' above elevation of pump house resulting in a pressure loss of abut 15 psi and minimum house pressure of 30 psi.

Operation:
      No. 1 — On 45 psi, off 60 psi
      No. 2 — On 40 psi, off 50 psi

Area lighting to be with 70 watt high pressure sodium vapor lights mounted on poles approximately 15 feet high. Space lights about 250 feet apart.

Water — Water mains to be cement-lined ductile iron, 6", class 52, tyton joints

Water — Water mains to be NSF approved PVC, SDR 21 for intended service.

Hydrants — Traffic type, open left, 1 - 4½" NST steamer nozzle and 2 – 2½" NST hose nozzles, 6" inlet mechanical joint connection, 5" valve seat, pentagon operating nut. Obtain approval from the local fire department.

Main Valves — Mechanical joint, resilient seat, butterfly, 2" operating nut.

Valve boxes — 5¼", screw type, cast iron lids marked "WATER"

Service line fittings — Mueller compression

Service lines — 1" type K soft copper unless otherwise noted

Restraints — 2500 psi thrust blocks plus integrally case restraint, or ¾" rod and friction clamps, or retainer gland

Main sewer to be 6" PVC, SDR 35. Sewer branches to each building to be 4" unless otherwise indicated.

Installation

Minimum slope for sanitary sewers shall be ¼ in. per foot unless otherwise indicated.

Where sanitary sewers cross water lines they shall be separated by at least 18". Whenever possible the water shall be above the sewer. For all crossings, keep sewer and water line joints equidistant with water joints at least 10' from sewer.

Water mains shall be flushed and disinfected with chlorine solution in accordance with AWWA C651-86. Have water tested by a recognized agency. Acceptable bacteriological sample must be collected before final health department approval for use of mains.

Test water supply system for at least one hour. Leakage shall not exceed that permitted by the AWWA equation:

$$L = N * D * P^{1/2} * 1850$$

where L = GPH; N = number of joints; D = diameter, inches; P = pressure, 50% above normal, 125 psi minimum.

After test, disinfect the water system by high pressure flushing the system and then applying a 100 ppm chlorine solution. The residual at the end of 24 hours at the system extremities shall not be less than 25 ppm. Thoroughly flush the system and apply normal pressures and disinfection.

Test sewers with deflectometer mandrel. Deflection shall not exceed 5%. Replace any pipe sections with excess deflection.

Test the sewer for leakage, not to exceed 20 gallons per day per inch pipe diameter per 1000 feet length. Sections shall be tested by filling manholes with at least 4 feet of water. Air testing of sewers will be considered. Submit details for approval.

Do not strip topsoil in sewerage system area or replacement system area. Stake and rope off sewerage system and replacement areas before start of *any* construction to prevent machinery access.

Septic tanks shall be placed a minimum of 10' from property lines and homes and 50' from wells. Absorption trenches shall be placed a minimum of 10' from property lines, 20' from homes, 35' from drainage ditches and 100' from wells and streams.

Work associated with the new entrance from NYS Route 96B shall be in accordance with NYS Department of Transportation specifications. Obtain permit from local office. Advise the local office of start of construction and obtain their continuing approval.

Immediately after grading of an area is completed, sow creeping red fescue at 60#/acre and annual ryegrass at 25#/acre for erosion control. Road-side ditches shall be particularly promptly treated.

Install straw bales in roadside ditches as grading is progressing. Reinstall as required at the end of each work day. Install new bales, minimum 100' intervals, after seeding.

Install rip-rap at all culverts and as directed.

All curves for roadways shall be elevated as directed by the municipality.

Install straw/hay silt fences in drainage ways to prevent erosion.

Soil Tests

Soil tests were performed by the Tompkins County Health Department.

Deep Holes

Test Hole	Depth Feet	Comments
1	6	Mottling at 18", evidence of high water
2	6	Slowly permeable soil, SPS, at 12", water at 3.5'
3	6½	SPS at 12", water at 4'
4	6½	SPS at 12"
5	6	SPS at 12", water at 4'
6	4	SPS at 12", water at 3'
7	4	SPS at 18", water at 4'

Perc Tests

Test Hole	Depth Inches	Percolation Rate-min/inch
1A	19	>60
1B	24	>60
3A	19	>60
3B	23	water
5A	18	water
5B	24	water
7A	18	water
7B	24	water

APPENDIX 10

Typical Realty Subdivision Drawings

Appendix 10 contains the set of drawings developed for a clustered housing realty subdivision. The homes are to be factory constructed modular housing.

Note: The drawings have been reduced to conform to page size. The format represented is important; the specific text is not.

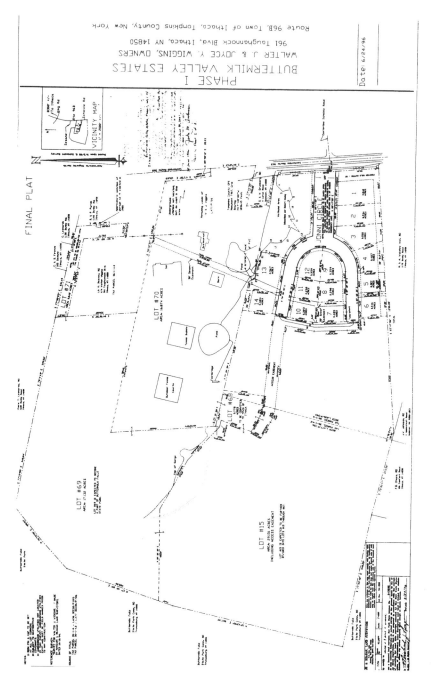

Figure 1
Surveyor's final plat.

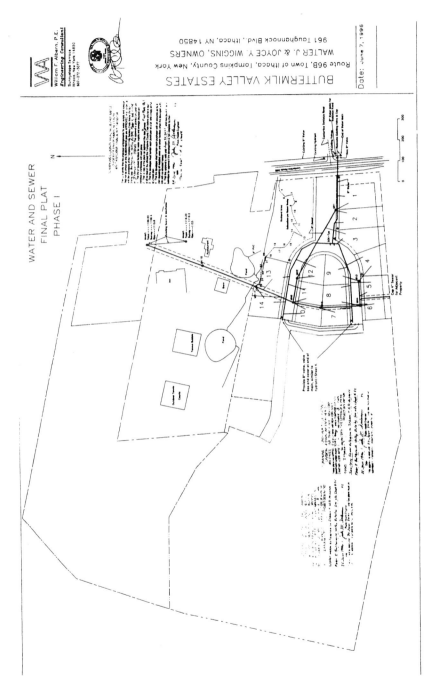

Figure 2
Water and sewer final plat.

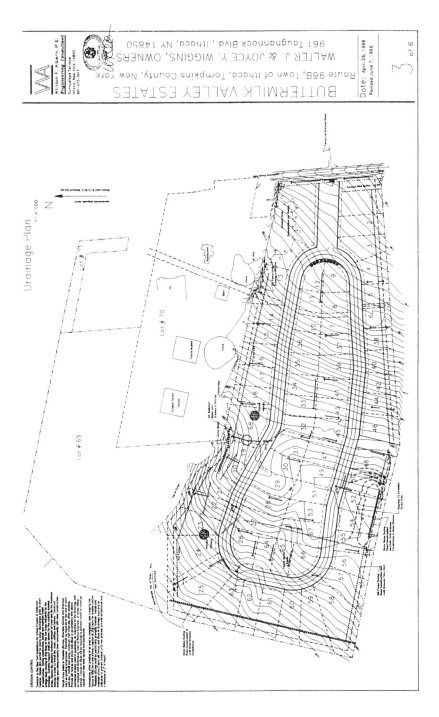

Figure 3
Drainage plan.

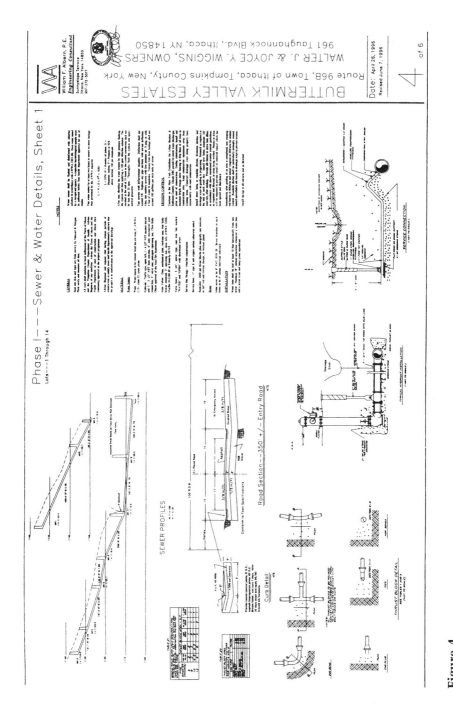

Figure 4
Phase I — sewer and water details, sheet 1.

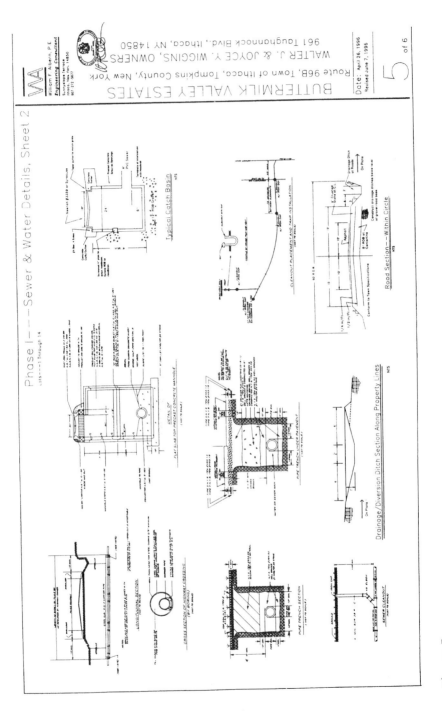

Figure 5
Phase I — sewer and water details, sheet 2.

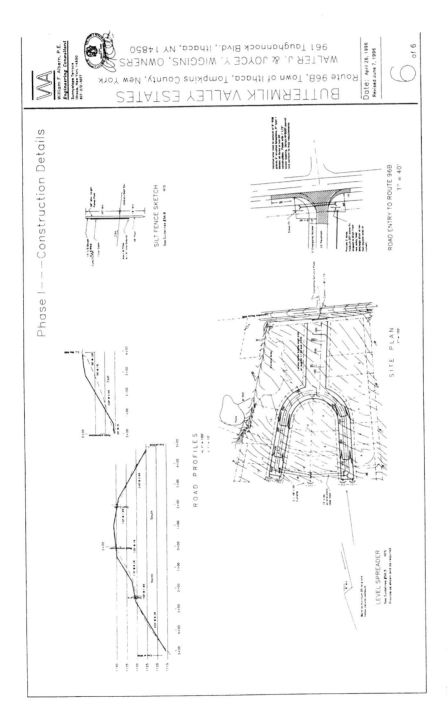

Figure 6
Phase I — construction details

APPENDIX 11

Typical Manufactured Home Community Drawings

Appendix 11 contains the set of drawings developed for a manufactured home community to serve an adult populace, over 55 year olds.

Note: The drawings have been reduced to conform to page size. The format represented is important; the specific text is not.

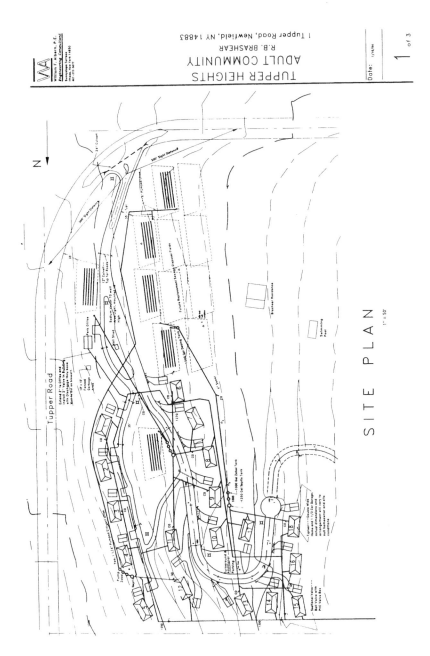

Figure 1
Site plan.

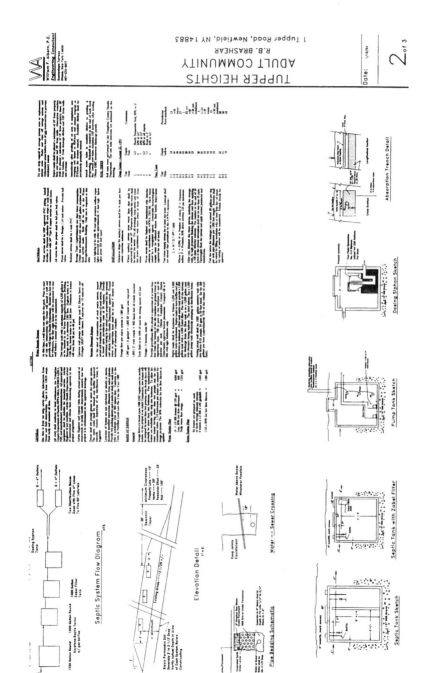

Figure 2
Details.

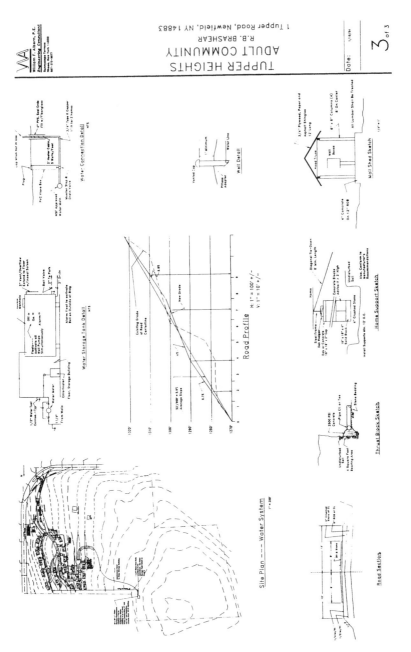

Figure 3
Details.

Index